T0293881

# God Exists

## New Light on Science and Creation

Joseph Davydov

**Schreiber Publishing**
Washington • New York

# God Exists
## by Joseph Davydov
### Published by:

## Schreiber Publishing, Inc.
P. O. Box 4193
Rockville, MD 20849 USA
spbooks@aol.com    www.schreibernet.com

### Library of Congress Cataloging-in-Publication Data

Davydov, Isai Shoulovich.
    [Sotvorenie i evolutsiia. English]
    God exists : new light on science and creation / Joseph Davydov.
        p. cm.
    ISBN 1-887563-51-2
        1. Creation. 2. Evolution--Religious aspects--Christianity. I. Title.

BS651 .D34 2000
231.7'652--dc21

00-024815

Printed in the United States of America

Everything that is good in me is primarily from God, and then from my father, teacher, and friend Shoul ben Matthew. To his bright honor this book is dedicated.

Joseph Davydov

And God created you, a human being, as a physical image of the nonphysical God, as a relative similarity of the absolute Creator. So you must deserve it and be grateful to your Creator. Love like God loves, and create like God creates. Do not kill, do not hurt, do not follow people of ill will, and to yourself be true. Remember that mortal bodies appear and disappear, while a deserving soul lives forever. Recognize the truth about the essence of existence in reliable scientific sources, and find out the purpose of your current life, where you came from and where you are going to. This book will help you better understand how scientific truth and Biblical truth can exist in harmony.

# About the Author

Joseph Davydov completed his PhD in 1967 at the Moscow Institute of Energy. In 1975, after the so-called "scientific atheism" of the former Soviet Union was forced to admit unconditionally the indisputable fact of the expansion of the Universe, it was "suggested" to instructors in all institutions of higher education in the Soviet Union that they promote "scientific" atheism as a preventive measure against "accidentally falling into religious delusion." Thus, in 1977, Davydov successfully graduated from the University of Marxism-Leninism in "scientific" atheism, whereby he conclusively confirmed his religio-scientific convictions.

After starting a new life in the United States, Davydov became free to reconcile his religious beliefs and his scientific knowledge. He received a professional engineer license in 1990 from the State of New York. He is now a full member of the New York Academy of Sciences (NYAS), and President of the International Scientific Center (ISC) in Brooklyn. He is the author of many theoretical works, such as "The Rational Solution of Unsolved Differential Equations with Periodic Coefficients," "The Theory of Mathematical Modeling of Dynamic Systems," "The Method of Damping or Exciting Vibrations and Oscillatory Processes," "The Theory of the Oscillating Universe," "The Theory of the Creation and Conservation of Energy," "The Theory of Multidimensional Space and of Communication Links between Physical and Nonphysical Worlds," and more. Davydov has published more than forty scientific works, including the book *Worlds*, which for the first time offers a proof of the objective existence of an Absolute God and other nonphysical worlds. The present volume was first published in Russian under the title *Creation and Evolution*.

# Contents

# Introduction

W e, the scientists of the former Soviet Union, were educated in the spirit of atheism. We were taught to idolize the Great Party and its leaders, Lenin and Stalin, and practice total intolerance of religious faith and God. In this lay the spiritual tragedy of the "Soviet People," especially its intellectuals, who were capable of critical thinking. Thank God a gradual recovery of this society is taking place at the present time, to the extent that it has become possible for us to familiarize ourselves with the true history of Bolshevism and to analyze critically the dogmas of the Soviet system, especially that which we called "scientific" atheism. In this context the book *God Exists* by Joseph Davydov is of great scientific and practical importance to everyone. Any unprejudiced reader who wishes to be exposed to authentic research will acquire a great deal of new and uncommon information. It is quite possible that for many it will become the first step toward a new "scientific-religious" world view, based not on blind faith but on scientific facts.

The first part of the Book, "God and the World," prepares the reader for a study of the Bible—a text given to us from above—in the light of science. Here the author acquaints us with the language of science, and does not shy away from serious scientific discussion of such subjects as the fundamental law of nature, the laws of the creation and conservation of matter, substance, and energy, the ideal and the physical, and the theory of the expansion and the evolution of the universe. Of particular interest is the chapter entitled "The Absolute God and the Relative World."

The second part of the book is devoted to an analysis of the six Biblical days of creation, or the six stages of the evolutionary development of the universe. By means of comparison the author succeeds in proving that advanced science does not negate but fully confirms the Bible. And this is his great merit.

Joseph Davydov continues the traditions of such authorities as Albert Einstein, Henri Poincaré, and Niels Bohr. The author possesses, not only a truly encyclopedic knowledge of the problem he examines, but he also makes masterful use of educational methods whereby the book becomes understandable to any reader, whether well schooled in the subject or not, young or old. The book is directed at the inquisitive person who desires to understand complex scientific data in common terms. I do not imply that the book can be read as a light or amusing novel. If, however, one reads it from cover to cover with sufficient attention, one will acquire the deep conviction that science and religion are two mutually complimentary aspects of our spiritual state, which never come into conflict as opposites.

Genrikh Golin
Professor of Physics, Doctor of Sciences,
President of the International Association
of Activists in Science and Culture

# Part One
# God
# and the
# World

We not only want to know how nature is constructed
(and how natural phenomena occur), but, if possible,
we would like to achieve a goal which might seem
utopian and arrogant at first glance, namely, to learn
why nature is what it is and not something else.

Albert Einstein

# =1=

## THE LANGUAGE OF THE BIBLE AND SCIENCE

*The stories in the Bible are more metaphorical than literal.*
Cyril Ponnamperuma

A theism is the opposite of religion. During the period of Stalinist persecutions in the former Soviet Union, atheism openly called itself "militant" and tried to suppress religion by brute force. The persecutions, however, were ineffective. They served to strengthen rather than undermine people's faith in God and the Bible. Hence atheism came to prefer the more effective tool of "persuasion" over coercion. At that point atheism proclaimed itself to be "scientific" and tried to use the unquestionable prestige of science to acquire legitimacy in the eyes of the masses.

Modern atheism has solemnly proclaimed itself "scientific," even though in reality it has not resolved any scientific problems but has merely proceeded from completely unfounded suppositions such as the "primacy of matter and the subordinacy of ideas." "Scientific" atheism has relied on such groundless and obviously false allegations to spread throughout the world the dogma that the hard facts of science discredit religion. This leads us to the important question of "What exactly is science?"

"Science is that realm of inquiry designed to produce new knowledge."[1] There are many natural sciences such as biology, chemistry, physics, cybernetics, astronomy, and so forth. "Philosophy" is the name given to that science which studies the essence of existence.

Modern philosophy is characterized by two opposite currents, namely idealism and materialism. Idealism is a school of philosophy which attempts to prove scientifically the primacy of objective ideas and the subordinacy of matter. For example, a person first thinks (or "gets the idea") of becoming a doctor, and only afterwards becomes one. Another example is the engineer's idea of designing a new airplane, following which an actual airplane is built. By perfect analogy, the idea of creating the world came first and only then was it followed by the actual creation of the Universe. Idealism is the philosophical foundation of religion. Materialism is the philosophical school which proceeds indiscriminately from the completely un-founded assumption of the primacy of matter and the subordinacy of ideas. All of the various forms of mechanistic and nondialectical materialism have discredited themselves so thoroughly that even atheism itself has completely abandoned them.[2] Hence the philo-sophical foundation of modern atheism is "dialectical" (or more precisely, Marxist) materialism.

If materialism, be it dialectical or nondialectical, were true, then the engineer would first build the airplane and only afterwards would think about building the airplane, writing the specifications, drafting the blueprints, and so forth. But it is impossible to build an airplane without blueprints, and it is impossible to produce blueprints without first thinking about them.

For many years, atheism has deliberately and methodically propounded the false dogma that the hard facts of the natural sciences disprove religion in general and the Bible in particular. It has preached this belief around the world so ardently and impressively that not just the wayward atheists and the doubters, but even some religious figures have come to believe it. As a result, the honest but inexperienced man is still troubled by the difficult and tormenting question of whether to believe the Bible or science. The atheists' presentation of scientific and technological advances has pushed people to believe in science, while life experience compels them to believe in the Bible.

Atheism has propounded this obvious falsity in order to prevent masses of people from seeing the truth in the Bible. Hence I believe that the question should be stated in a completely different way, i.e., "Does modern science really discredit the Bible or does it in fact corroborate it?"

This question can only be answered correctly by comparing the verses of the Bible with the corresponding facts of modern science dispassionately and objectively. Here I will limit myself to a comparison of this evidence for the first book of the Bible, Genesis, which is devoted to the most fascinating problem of the creation and evolution of the Universe. In doing so, I must stipulate that the results of my analysis and comparison could not possibly be objective and dispassionate unless I took into account the peculiarities of the languages in which the documents in question were written. I know that identical semantic contents may be written in different languages: English, Russian, Hebrew, Chinese, Farsi, Turkish, and so forth. One can only compare texts written in different languages after first translating all of them into a common, mutually comprehensible language. Every language has its own unique history. For example, Middle English is markedly different from Modern English. An American textbook on the history of the English language gives the following Middle English text (1300 A.D.), its modern English translation, and a comparative analysis as an example to illustrate the changes which have taken place.[3]

### Northern

*The Cursor Mundi*, c. 1300.

> Pis are pe maters redde on raw
> Pat i thynk in pis bok to draw,
> Schortly rimand on pe dede,
> For mani er pai her-of to spede.

Notful me thinc it ware to man
To knaw him self how he began, --
How [he] began in werld to brede,
How his oxspring began to sprede.
Bath o þe first and o þe last.
In quatking curs þis world es past.
Efter haly kyre[es] state
Þis ilk bok it es translate

In to Inglis tong to rede
For þe love of Inglis lede,
Inglis lede of Ingland,
For þe commun at understand.
Frankis rimes here I redd,
Comunlik in ilk[a] sted:
Mast es it wroght for frankis man.
Quat is for him na frankls can?
Of Ingland þe nacion--
Es Inglis man þar in commun--
Þe speche þat man wit mast may spede.
Mast þar-wot to speke war nede.
Selden was for ani chance
Praised Inglis tong-in france.
Give we ilkan þare language,
Me think we do þam non outrage.
To land and Inglis man i spell
Þat understandes þat i tell....

TRANSLATION: These are the matters explained in a row that
I think in this book to draw shortly riming in the doing, for many
are they who can profit thereby. Methinks it were useful to man to
know himself, how he began, how he began to breed in the world,
how his offspring began to spread, both first and last, through what
kind of course this world has passed. After Holy Church's state this
same book is translated into the English tongue to read, for the love

of English people, English people of England, for the commons to understand. French rimes I commonly hear read in every place: most is it wrought for Frenchmen. What is there for him who knows no French? Concerning England the nation—the Englishman is common therein—the speech that man may speed most with, it were most need to speak therewith. Seldom was by any chance English tongue praised in France. Let us give each their language: methinks we do them no outrage. To laymen and Englishmen I speak, that understand what I tell.

You would hardly be able to understand this Middle English text without some preliminary explanation or decoding, even if you had an excellent command of modern English. But after all, this text was written "just" 700 years ago. Moses wrote the Bible 3300 years ago. Mere centuries, not to mention millennia, are capable of altering a language beyond recognition. On this basis, just imagine the extent to which the ancient Hebrew language has changed since then and what kind of explication or decoding is required for the Bible, which was written not 700, but 3300 years ago.

Moreover, words and phrases with the same semantic contents written in the same national language may differ dramatically from one another depending on their purpose, i.e., whether they are intended for a specialist in a particular field, a general audience, or children. Because of this, one must first distinguish between science, popular science, and children's literature. Children or semiliterate adults are not the only ones who cannot understand scientific literature. Even scientists in unrelated fields might have a hard time understanding it. If one wants to bring the results of an objective comparison of ancient Biblical and modern scientific materials to the attention of the general public, one must first translate both sets of materials into the easily understandable popular scientific language of today. Without this translation and without this kind of preliminary explication of the Bible, it would be impossible to speak of an objective solution to the problem at hand. Therefore, instead of comparing the vernacular presentation of the modern scientific model of the evolution of the Universe with the literal version of the Biblical model of the creation, I will use a popular scientific (decoded) translation of the Biblical model.

The Bible was not written in scientific language and was not written for modern scientists. It was written in the ancient Hebrew vernacular for the common people. That is why the masses have been able to understand and appreciate it for 3300 years. No other book is comparable to it, either in terms of the profundity of its meaning or the facility of its presentation. The ancient Hebrew language, i.e., the language of slaves whom the Pharaohs forced to mix mortar and build the pyramids 24 hours a day, did not contain such philosophical concepts as idea, matter, substance, energy, molecule, atom, proton, electron, hydrogen, plasma, planet, stage, era, period, and so forth. Hence, in the Bible, depending on the context, the same common word could and had to express a variety of scientific concepts. For example, the word "earth" not only expresses the concept of land but also conveys the concepts of planet, material, matter, the galaxy, the Universe, and the Material World.

Criticizing the Bible because it names the Material World "earth" and the Ideal World (i.e., the opposite of the Material World) "heaven" is equivalent to criticizing the English language as unscientific because it sometimes uses the word "world" instead of the word "globe."

If you say that someone has traveled "around the world" instead of "around the globe," no one would even think of reproaching you for it. Nevertheless, well trained and highly educated atheists are past masters at intentionally confusing similar concepts and using these "word games" to their own subjective and unscientific atheistic advantage.

I should point out that even in modern English the word "earth" expresses four different concepts:

1) The land, as distinguished from sea or sky.
2) The planet we live on.
3) This World or The Temporal World, as distinguished from another (nonphysical) world.
4) Soil, ground.

The word "idea" in modern English expresses five different concepts:

1) Opposite of matter.
2) A thought, mental conception, or image.
3) An opinion or belief.
4) A plan or scheme.
5) The meaning or significance.

The word "world" in modern English has seven meanings:

1) The total system of all Universes.
2) The Universe.
3) The earth.
4) Some part of the earth (Old World).
5) The public.
6) Any sphere or domain (the cat world).
7) Individual outlook (his world is narrow).

The word "light" in modern English has four meanings:

1) The form of radiant energy.
2) Public view (to bring new facts to light).
3) Having little weight.
4) Easy to do (light work).

Even in modern English, the word "day" has five meanings:

1)  Period of light between sunrise and sunset.
2)  The time (24 hours) that it takes the earth to revolve once on its axis.
3)  A period or era.
4)  A time of power, glory, etc.
5)  Daily work period (7 or 8 hours).

I could cite a multitude of such examples for any language.

The word "water" in the Bible expresses the concepts of not just water ($H_2O$) but also hydrogen, hydrogen plasma, and hydrogen cloud. Thus, for example, the expression "the waters which were under the firmament"[4] refers to the cloud of hydrogen plasma from which our solar system subsequently originated. The expression "the waters which were above the firmament" refers to that cloud of hydrogen plasma from which all the other stars and galaxies formed. The expression "And the Spirit of God moved above the face of the waters" may be translated into modern scientific language as something like this: "the ideal (nonmaterial) program for the evolution of matter created by God was encoded in the newborn Universe before the appearance of the clouds of hydrogen plasma which were subsequently transformed into the galaxies and stars."

According to the Bible, God is an *ideal*, and not a *material* category. This means that God does not have a material body, a material tongue, or material eyes. Hence the Biblical expressions "And God said" as well as "and God saw" should not be taken literally. After all, we don't interpret the English word "saw" literally when we say "I saw you in a dream" (a person's eyes are closed when he or she is sleeping and therefore he or she cannot "see" anything with his or her eyes). Another example: if two people are listening to the radio with their *ears* instead of their *eyes*, even so one of them might ask the other quite properly: "You see what's happening, don't you?"

In the words of the American scientist Cyril Ponnamperuma, the stories in the Bible "are more metaphorical than literal."[5] And when the Bible talks about God, then the words "said" and "saw" are figurative, metaphorical expressions, not literal. Hence the Biblical phrase "And God said" in no way means that God uttered any kind of audible speech in English, Hebrew, or Russian, but instead merely expresses the intention to carry out a plan. The Biblical phrase "And God saw that it was good" shows God's conviction that each stage of the process was consistent with the overall program for the creation of the world and was an essential link in the chain of God's creativity. In modern popular scientific terms, this might mean something like this: "and God was assured that the ideal program was encoded in the end results of the current phase of evolution and that all of the favorable conditions necessary for the next phase to begin were embedded in the end results." I should point out that the language of the Bible is easily understood by all people, both ancient and modern. If the Bible had been written in modern popular scientific language, people during the nineteenth century would not have been able to understand it, let alone ancient people.

For the sake of objectivity, I should mention G. Sinaysky's comments on the characteristics of the ancient Bible and modern science:

> The Bible does not contain a comprehensive system of the Universe (cosmology), nor does it contain a complete and comprehensive doctrine of the origins of the world and life (cosmogony). The Bible merely provides a concise account of the beginning which can be understood by even the most uneducated, even semiliterate man, but obviously requires explication.

Conversely, science holds extremely uncertain and contradictory notions of the beginnings, has more or less clear concepts of
the development of the world, but knows absolutely nothing about
the end of history. In other words, science doesn't know where it's
going! The more a scientific genius knows about the world right
now, the more mysteries and uncertainties there are. Science
studies individual aspects of existence in isolation from the cosmos
as a whole."[6]

"Astronomy is the science studying the location, motion,
structure, and development of celestial bodies, their systems, and
other forms of cosmic matter."[7] Astronomy consists of the following
disciplines: astrophysics, cosmogony, cosmology, and so forth.
*Astrophysics* studies the physical properties of cosmic matter and fields,
while *cosmogony* studies their origin and development. *Cosmology* is a
science which studies "the general characteristics of the structure of
the Universe as an integral coherent whole and as an all-encompass-
ing system of cosmic systems."[8]

In one sense, Sinaysky is right. But when he carelessly asserts that
the Bible contains absolutely no comprehensive doctrine concerning
the structure and origins of the world, I will be so bold as to respond
in the following way: if one were to summarize the entire scientific
content of modern astronomy (astrophysics, cosmogony, cosmology,
and so forth) in a few pages, in very plain language which could be
understood by even the most unlearned or even illiterate person, then
one would inevitably wind up with the first book of the Bible!

The genius of the Bible primarily lies in the fact that it is written
in a concise and plain language which can be understood in any era,
by any nation, and by all kinds of people, regardless of their level of
intellectual development. It is so condensed that it is not amenable to
scientific analysis in just one or even several books. Hence here I will
examine only the first (and most fascinating!) book of the Bible,
which concerns the creation of the Material World by the Ideal God.

In the first part of our book I shall introduce the general fundamentals of modern science to the reader, while in the second part I will examine and compare the scientific and Biblical models of the six stages of the creation and evolution of the Universe.

While scientists in the West can write about their religious beliefs openly and candidly, in pre-Glasnost Soviet-bloc scientific literature these beliefs could only be read between the lines, and not in the lines themselves. In this sense an analysis of scientific writing published in the Soviet Union prior to Gorbachev's accession to power would be especially valuable, because it would convincingly demonstrate that not only free science, but even science which was systematically controlled by totalitarian atheism, in its essence and its content corroborates religion instead of discrediting it. It would show that any objection to religion is politically motivated and unscientific. In order to convince the reader of this, I will mostly cite Soviet scientific literature published before 1985.

An analysis of this literature convincingly demonstrates that the power of the Bible lies in its accuracy. *The Bible is powerful because it is true.*

The triumph of scientific religion over totalitarian atheism was no coincidence and did not occur because of the collapse of the powerful Communist Party of the Soviet Union, but because religion contains the truth.

It was not atheism that collapsed as a result of the disintegration of the Communist superpower, it was the Communist superpower which collapsed because of the utter scientific bankruptcy of totalitarian atheism.

The Communist superpower could not have built up its military might without the development of natural sciences such as physics and astronomy. The development of the natural sciences was inevitably accompanied by the triumph of truth over falsehood, meaning that atheism lost its foothold little by little. Ultimately "scientific" atheism completely succumbed because of its total scientific failure at a time when the military might of the Communist superpower had reached the apex of its development. "Scientific" atheism was the foundation of "dialectical" materialism, and "dialectical" materialism was the basic ideology ("soul") of the Communist Party of the Soviet Union. The Communist Party of the Soviet Union (as a political force) lost its basic foundations, its ideology, and its "soul," and therefore suffered a total defeat, despite its military (physical!) power.

The "ideas of Communism" were the first to emerge, and only afterwards was the Communist superpower created.

The decline of Communism followed the same basic pattern: the scientific collapse of the "ideas of atheism and Communism" came first, followed by the destruction of the physical power of the Communist Party of the Soviet Union. Thus, in spite of its unfounded assertions, the course of development of the Communist Party of the Soviet Union provided a demonstration of the *primacy of the idea and the subordinacy of matter* for the entire world.

# =2=

# SUBSTANCE AND ENERGY

*Inert mass is nothing more than latent energy.*
Albert Einstein

The ancient Hebrew language of the Bible did not contain the commonly accepted philosophical and scientific concepts of today, such as idea, matter, substance, energy, molecule, atom, nucleon, proton, neutron, electron, hydrogen, and so forth. Not only were ancient people unfamiliar with these words and concepts, but so are many present-day people. Therefore, I will introduce the reader to the definitions of several of the basic terms used in modern science.

The natural sciences use the concept of *mass* to describe the amount of physical reality contained in a particular object. According to the Theory of Relativity, the mass of a body increases with an increase in the velocity of its motion. The mass of a body is lowest when the body is at rest. This minimum mass of a body is known as *rest mass*. Weight is that force which compels a body to fall to the earth (or any other planet). Weight is equal to the product of mass by free fall acceleration, which varies from planet to planet. For example, the acceleration of a free-falling body on earth is equal to 981 cm/sec$^2$. Therefore, the exact same physical body will have different weights on different planets, even though its rest mass will remain the same in any system of reference. If weight is expressed in kilograms, then mass should be expressed in terms of kg times sec$^2$/m. For the sake of simplicity, however, we will often express rest mass in kilograms, thus arbitrarily equating it to a proportional weight.

I will call any category whose rest mass is not equal to zero *weighty*.

A weighty physical body will not cease to be weighty even if its free fall acceleration equals zero in certain situations. If free fall acceleration becomes equal to zero in a certain place or at a certain time, then one would say that the *weighty body is in a state of weightlessness*. Therefore, when I use the term *weightless* without further qualification (as in "state of weightlessness"), I will be referring to bodies whose rest mass is equal to zero.

I will use the term substance[1] to refer to any material which possesses weight and physical volume. *Rest mass* is the quantity which characterizes the amount of substance contained in a particular body. All of the bodies, objects, and things around us are constructed from different substances. For example, a desk consists of wood with a small amount of metal parts. A window consists of glass and a plastic frame. Tea consists of water and sugar. In this case wood, glass, plastic, water, sugar, and so forth have a common name, namely *substance.*

Any substance necessarily possesses weight, volume, and specific physical and chemical properties which will not change in the aggregate as a result of simple division. Even though any part is less than the whole, the total weight and total volume of all the parts will always remain equal to the weight and volume of the whole. To put it in simpler terms, a substance which undergoes simple division will remain the same substance.

For example, if we saw a piece of wood into several parts, then the wood will remain wood regardless, because none of its physical or chemical properties will change.

If we take any of the smaller pieces of wood and saw it into even smaller pieces, then once again the wood will still be wood.

However, we could not continue these kinds of simple divisions indefinitely without causing a change in quality. According to the dialectic law of the transformation of quantitative changes into qualitative changes, there exists some definite number of divisions after which wood will cease to be wood.

Here is another example. If we pour a large glass of water into small glasses and then divide the contents of these smaller glasses into drops, and divide each drop into still smaller drops, then this would not cause this water to stop being water. But we cannot continue these divisions indefinitely without causing a change in quality. According to the dialectic law of the transformation of quantitative changes into qualitative changes, there exists some definite number of divisions after which water will no longer be water. The smallest particle of water which cannot be divided without causing water to stop being water is called a *molecule* of water. If we continue to divide this smallest particle into its component parts, then instead of a particle of water we would get the smallest particles (molecules) of two other substances, namely hydrogen and oxygen, whose chemical properties are very different from the chemical properties of water.

In the examples cited above, the substance in question does not possess any charge. We will use the term *neutral* to refer to those types of substances which do not possess any charge.

A neutral particle of a weighty and visible substance which further division would be impossible without changing its chemical properties is called a *molecule*. Every molecule always has some elementary weight and some elementary volume. That is why it remains a substance. But the volumes and weights of molecules are so small that they cannot be seen with the naked eye. A molecule is a particle of a visible substance which is invisible to the naked eye. And if we cannot see molecules with our own eyes, it does not in any way mean that they do not exist. Molecules exist objectively and independently of the idiosyncrasies of our vision.

A molecule cannot be divided into its constituent parts without undergoing a change in chemical properties. But this does not imply that a molecule is some kind of indivisible particle. A "large" molecule can always be divided into small molecules, and small molecules can always be divided into even smaller molecules, and so forth. Despite this division of molecules, the substance still remains a substance. It merely undergoes a change in chemical properties, meaning that some kinds of substances are transformed into other kinds of substances.

But this kind of division of molecules cannot be continued indefinitely without causing the molecule to cease being a molecule. According to the dialectic law of the transformation of quantitative changes into qualitative changes, there is a definite limit at which a molecule can be divided only into its weighty "building blocks," which are known as *atoms*.

An *atom* is a part of a molecule which is also a kind of microscopic model of the colossal solar system (the radius of an atom is $10^{23}$ times smaller than the radius of the solar system). While the solar system consists of the hot sun at the center and cold planets in orbit around it, the tiny atom consists of a central positively charged nucleus and negatively charged electrons orbiting around it. The algebraic sum of all the electrical charges of an atom is always equal to zero, so that the atom as a whole is an electrically neutral weighty particle.

While molecules are invisible to the naked eye, their smallest particles, i.e. atoms, are far more invisible. If, for purposes of comparison, we put a metal ball 1 mm in diameter in our hands, then the diameter of the atom is 100 million times smaller, and the radius of the atom's nucleus is 10,000 times smaller than the radius of the atom itself. The radius of the sun is smaller than the radius of the solar system by approximately the same amount. This kind of similarity demonstrates that the gigantic solar system and the tiny atom, which is invisible to the naked eye, are different details of the same creative design or different products of the creativity of one and the same intelligent creator.

A molecule consists of some set of identical or different atoms. For example, a hydrogen molecule consists of two atoms of hydrogen, while a water molecule consists of two atoms of hydrogen and one atom of oxygen. "Molecules of the most complex substances, i.e. the higher proteins and nucleic acids, are constructed from hundreds of thousands of atoms. In the process atoms may be combined with one another in different ways as well as in different proportions. Hence, a relatively small number of chemical elements can produce an extremely large number of different substances."[2]

The natural sciences have definitely proven that all of the things and objects around us, i.e. solids, liquids, gases, the stars, the planets, plants, life forms, and so forth, consist either of molecules or elementary particles and combinations thereof, which under certain conditions may combine into molecules. The term *"elementary"* usually refers to *particles* whose internal structure cannot be represented as a combination of other particles at the current stage of development of physics.

As applied to particles, the concept of "elementary" is not just a relative but an arbitrary category, because over time physics has penetrated deeper and deeper into particles which were previously considered elementary. Atoms are not elementary particles in that they consist of even smaller components such as electrons, protons, and neutrons.

An *electron* is an elementary particle which possesses a rest mass of 9.1 times $10^{-31}$ kilograms and an elementary charge of electricity. In this case an *elementary charge of negative electricity* means the smallest negative electrical charge which exists in nature, (according to current conceptions), i.e., $e = 1.602$ times $10^{-19}$ coulombs. The charge of any body is an integer multiple of $e$. The weight of an electron is almost 2000 times less than that of the simplest hydrogen atom.

The nuclei of atoms consist of protons or combinations of protons and neutrons. A *proton* is an elementary particle which has a rest mass of 1.67265 times $10^{-27}$ kg and a positive electrical charge, which in terms of magnitude value is equal to an electron charge. A *neutron* is an electrically neutral elementary particle with a rest mass of 1.67495 times $10^{-27}$ kg. Scientists are still referring to protons and neutrons as elementary particles, despite the fact that modern physics already has some evidence of their internal structure.[3] That is why we must once again emphasize the arbitrary and relative nature of the concept of "elementary" as it applies to particles. Protons and neutrons are referred to in common as *nucleons*.

Thus, the natural sciences have definitely proven that all the objects around us consist of molecules.[4] In turn, all molecules consist of atoms, and atoms consist of electrons and nucleons (protons and neutrons).

In our daily lives we can clearly see that we are surrounded by weighty and visible physical bodies. The larger the body, the more noticeable it is to us and the better we see it. Hence we get the deceptive impression that large bodies are somehow more significant than small bodies. But in fact this in no way implies that weightless and invisible reality is somehow nonexistent or that weightless and invisible categories are somehow less important than weighty and visible categories.

While we can see large physical bodies with the naked eye, we can only see the small molecules which make up these bodies with the aid of a microscope. Nevertheless, the structure of the desk at which we sit is far simpler and more primitive than the structures of the molecules which make up the desk and which we cannot see because of their extremely small size. The structures of the tiny molecules which cannot be seen with the naked eye are not necessarily simpler and may be far more complex and efficient than the structures of large, heavy bodies.

For example, small and complex DNA molecules are far more sophisticated than large and simple bodies. If we were to assign ourselves the task of transcribing the entire contents of the genetic program encoded in the DNA molecules which are "packed" in the nucleus of a living cell, we would have to write so many thick books that we would have a hard time accommodating them in any major library. The structure of a DNA molecule, which is invisible to the naked eye, is much more sophisticated than the structure of the body of a huge elephant or whale. In this case the small molecule acts as a complex limit toward which the simple form of a large organism tends when we mentally and systematically divide it into its constituents.

If we divide any molecule into individual atoms, then each atom after the division, just like each molecule before the division, will have specific dimensions, weight, and rest mass. If we then proceed to divide the atom into nucleons and electrons, then each nucleon and each electron after division, just like each atom before the division, will have specific dimensions, weight, and rest mass. But this kind of division cannot go on indefinitely without altering the quality of weight. According to the dialectic law of the transformation of quantity into quality, after a specific number of divisions, weighty, elementary particles possessing a specific volume and rest mass will be transformed into pure energy with absolutely no physical volume, no weight, and no rest mass.

According to the classical definition, *energy* is a general measure of physical work capacity. According to Einstein's Theory of Relativity, energy is proportional to mass and is therefore a quantitative measure of substance also. Elementary portions of pure solar energy are commonly referred to as *photons*. The physical volume, weight, and rest mass of a photon are equal to ideal zero. *The photon is a persuasive example which demonstrates that not every objective reality has weight or physical volume.* The proof of this proposition is extremely simple.

Photons are propagated in a spatial vacuum at the extremely high (scientists believe the maximum possible in the Universe) velocity of "c," which is equal to 299,792 kilometers per second. Any other velocity would be unacceptable, which means that it would be absolutely impossible to accelerate a photon. Weight is equal to the product of mass by acceleration, and the acceleration of a photon is always equal to zero. Therefore a photon's weight is also equal to zero and the photon has no weight whatsoever.

A photon can never be anything but a photon. It is impossible to stop, or even reduce the speed of a photon, which constantly moves at the speed of light. If you try to do this, the photon will cease to be a photon. Thus, we cannot even talk about the rest mass of a photon.

From physics we know that any weightless energy can only exist in the form of a wave (electromagnetic, light, biological, and so forth). And an unlimited number of waves can pass through an infinitesimally small volume of physical space simultaneously. This means that the volume of an elementary portion of any weightless energy is equal to zero.

"Scientific" atheism and "dialectical" materialism refer to both weightless physical energy and weighty substances by the common name of *matter*.[5] We will not object, because the name does not change the matter one bit. However, we will try to convince the reader that "scientific" atheism and "dialectical" materialism cannot save themselves from total scientific disaster even by calling pure and weightless energy matter.

Thus, matter may be weighty or weightless. Weighty matter is commonly called substance, while weightless matter is called either energy or a field. A physical field exists in the form of weightless waves whose elementary portions are in a state of constant motion at the so-called velocity of light, which for a perfect vacuum equals 299,792 km/sec. The natural sciences have definitely proven that substance is a weighty form of weightless energy.[6] According to the Special Theory of Relativity, the amount of energy concentrated in a substantial body may be expressed by the formula:

$$E - m_r c^2$$

where $m_r$ is the rest mass of the body and c is the velocity of light in a perfect vacuum. Thus, the basic material and "building blocks" for the amazing variety of all weighty (i.e., substantial) elements and systems is weightless positive energy.[7] All objects, molecules, atoms, nucleons, and electrons have their own intrinsic dimensions, weight, and rest mass and are therefore *substantial* categories.

Moreover, matter may be visible or invisible. Convincing examples of the existence of invisible reality are provided by *radio waves*, which everyone is familiar with and which freely travel through the thick walls of buildings but which one cannot touch, see, hear, and so forth. You can't listen to the radio waves themselves, only to the sound waves into which the radio waves are converted by the receiver. Although we can't see, hear, sense, or touch radio waves, this by no means implies that they do not exist. A physical field of invisible, inaudible, imperceptible, and untouchable energy waves exists objectively and independently of our bodily capacities. The air which we breathe but cannot see is a very simple example of invisible reality.

The structure of photons, which have no weight and no volume, is much more complex and efficient than the structures of atoms and molecules which do have an elementary volume and weight. While we might call atoms and molecules "complex" by comparison with large and simple objects, we should call the same atoms and molecules "simple" by comparison with weightless photons, whose volume is equal to zero. In this case the form of the photon serves as the complex limit toward which the simple form of a substantial electron tends when we mentally and systematically divide it.

According to the theory of the Soviet academician M. A. Markov, modern scientists conceive of the photon as a unique "microworld" with a unique "microcivilization," even though its physical volume is equal to ideal zero.[8]

We may arrive at similar conclusions when we examine any other aspect of the world around us. Then, by using the scientific method of induction and going from particular examples to the general truth, we may formulate the following conclusions:

1. We call any purely physical phenomenon and any purely

physical reality a material category. Matter may be weighty or weightless, visible or invisible. We call weighty matter substance and weightless matter pure physical energy. Pure energy which is weightless under any conditions should be clearly distinguished from weighty substance even when this weighty substance is in a state of weightlessness.

According to the Theory of Relativity, energy is a measure not only of substance but of its physical work capacity. As a measure of substance, energy is the initial material and "building block" for the amazing variety of all substantial bodies. Any substance is a weighty manifestation of weightless positive energy. As a measure of work capacity, energy is responsible for the motion and development of material elements and systems of the physical world.

2. The structure of the tiniest particles of matter, which are invisible to the naked eye, is not supposed to be simpler, but might be much more complex and efficient than the structure of large and weighty bodies.

3. Weighty and visible substance is the crudest form of objective reality. Weightless energy is objective reality of a higher and more refined quality than weighty and visible substance which has volume.

4. Not every physical reality is weighty and visible. Only the crudest form of physical reality, namely weighty substance, possesses physical volume. The physical volume of any other category of objective reality is equal to ideal or absolute zero.

Besides its whims and illusions, what does atheism have in its arsenal to counter these scientific conclusions? The only thing it has to oppose these scientific facts is its completely unproven basic assumptions that "the world contains nothing except for weighty and visible matter."

But then we might rightly ask where the scientific proofs of atheism are. The answer is very simple: scientific atheism never had any scientific proofs and still doesn't have any! It only has its convenient initial assumptions, which hundreds of millions of people are still compelled to believe blindly right now, even though they are logically impossible, because the natural sciences have definitely proven that weightless photons with no volume really and truly do exist. If there were no solar energy photons, you or I would not be here either. Moreover, everyone (even the most illiterate person!) knows from his or her own experience of the existence of weightless and invisible radio waves which penetrate thick brick or concrete walls to reach one's receiver.

# =3=

## PARTICLES AND ANTIPARTICLES

*A particle only occurs in a pair with its antiparticle.*
Vitaly Rydnik

The philosophical foundation of modern atheism is dialectical materialism, which proceeds without any proof whatsoever from the notorious assumption that "there is nothing in the world except for matter in motion." But then we might ask the completely valid question of what does atheism mean by this absolutist concept of matter.

essential characteristics of a particular object (element or system) are known as its *attributes*. If an attribute is lost, then the object itself ceases to exist. So what would be an essential property of matter without which matter would cease to be matter? Previously, materialists recognized the following universal attributes of matter: volume, dimensions, length, width, height, weight, mass, and so forth. In reality, all solids, liquids, and gases consist of molecules. All living organisms also consist of molecules. In turn, molecules consist of atoms. Atoms consist of electrons, protons, and neutrons, which have all of the aforementioned attributes.

But scientific developments prepared unexpected surprises for the materialists. For example, in 1900 the German physicist Max Planck discovered that light is propagated, radiated, and absorbed discretely, in fixed portions known as *quanta*. Light energy *quanta* later became known as *photons*. The fact that a photon has no extension, no volume, and no weight whatsoever is extremely interesting.

All of the photon's dimensions, volume, length, width, height, and weight are equal to ideal zero. These discoveries were positively shocking for the materialists. After all, they led to the obvious conclusion that the "building blocks" of matter consist of nonmaterial energy. And this means, that the foundation on which everything around us stands is a nonmaterial one, i.e., that matter is secondary to nonmaterial energy. Hence it would have been quite logical if atheism had exited the world arena back then. But instead, atheism began to adapt its basic dogmas to the latest scientific advances. To do so, it first had to significantly expand the definition of matter. First of all, atheism solemnly declared that even pure energy, including the photon, is also matter. Secondly, the atheists asserted that dimensions, volume, weight, and rest mass are not universal attributes of matter but only belong to that type of matter known as substance. They then blamed so-called mechanistic (i.e., nondialectical) materialism for equating the concepts of matter and substance.

Before atheism could completely recover from this surprise, scientific advances prepared another one. Paul Dirac, the British physicist and founder of relativistic quantum theory, theoretically predicted the objective existence of elementary antiparticles and antisubstance constructed from it.

An elementary antiparticle was first discovered experimentally in 1932. This was a positron captured on film. Approximately 20 years later experimenters photographed an antiproton, the electrical opposite of the proton.[1]

This compelled "dialectical" materialists to adapt their dogmas once again to scientific advances and expand the definition of matter even more. According to their new concept, atheism recognizes three basic kinds of matter: substance, antisubstance, and the field.

In physics a *field* is a physical continuum which has no rest mass (electromagnetic field, light field, gravity field, biological field, and so forth). From this definition it is obvious that the scientific and philosophical meaning of "field" differs markedly from its everyday meaning. While a field in physics usually means a three-dimensional region of a continuum or space, in daily life it means a two-dimensional tract of land on which something is planted.

In contrast to a field, *substance* is the term used to describe an aggregate of discrete concentrates of positive energy with a positive rest mass (election, protons, neutrons, atoms, molecules, and everything constructed from them). "Dialectical" materialism and "scientific" atheism have admitted that substance could not exist without its opposite, i.e., without antisubstance.

*Antisubstance* refers to the electrical, energetical, or any other opposite of substance which has a rest mass (antielectrons, antiprotons, anti-neutrons, antiatoms, antimolecules, and everything constructed from them). We will use the term *electro-antisubstance* to refer to the electrical opposite of substance which possesses a positive rest mass. We will use the term *energo-antisubstance* to refer to the energetical opposite of substance which has a negative rest mass. Usually the literature refers to energo-anti-substance as antimatter. But atheism uses the term "energo-anti-substance" to refer to a mere variety of matter and not its opposite, even though it calls it antimatter.

However, names will not change the essence. In any case, if substance is a weighty manifestation of weightless positive energy, the energo-anti-substance (or antimatter) is a weighty manifestation of weightless negative energy. When substance is combined with its electrical opposite, both lose their rest mass and are completely transformed into weightless positive energy; however, when substance is combined with its energetical opposite, the positive energy of the substance and the negative energy of the antisubstance completely destroy each other, leaving no matter whatsoever.

It would be practically impossible to detect energo-antisubstance. If we tried to come into contact with it, it would prove fatal to us, because the positive energy of our physical bodies and the negative energy of the energo-antisubstance would completely destroy each other without being transformed into anything material.

In the view of modern-day physics, substance should have a large number of other opposites, both material and nonmaterial. In this case *opposites* should be understood as objectively existing elements or systems which are equivalent to one another in magnitude (in quantity) and opposite in sign (in quality).

We should point out that opposites may be classified as component or dialectical. Componential opposites mean opposites which must exist simultaneously. *Componential* opposites should not alternate over time. In this case the most important difference between them and dialectical opposites, which must periodically alternate and follow one another over time, is like the difference between night and day. One and the same opposites may be both componential and dialectical (for example, light and dark), but not necessarily. Componential opposites are indispensable constituents of any matter as a physical reality.

Any substance is constructed from particles with an elementary rest mass. These weighty particles are in turn formed from such weightless elementary portions (particles) of pure energy such as photons whose rest mass is equal to ideal zero.

At present we know about more than one hundred different types of elementary particles. The most interesting are photons, neutrinos, electrons, protons, gravitons, and so forth. Every elementary particle has an equivalent opposite, which we call an *elementary antiparticle.* Elementary antiparticles respectively include antiphotons, anti-neutrinos, positrons, antielectrons, antiprotons, antigravitons, and so forth.

Like particles, antiparticles may be weighty or weightless. Weightless antiparticles (such as the antiphotons of physical space) are portions of negative energy. Weighty electro-antiparticles (such as positrons and antiprotons) are formed from weightless particles in the form of weighty manifestations of weightless positive energy. Weighty energo-antiparticles (such as antielectrons and energo-antiprotons) may be formed from weightless antiparticles in the form of weighty manifestations of weightless negative energy.

The "building blocks" of any electro-antisubstance, like substance, consist of weightless positive energy, while the "building blocks" of energo-antisubstance consist of negative energy. Electro-antisubstance is constructed from electro-antiparticles, while energo-antisubstance is constructed from energo-antiparticles.

"*Positron*" is the term used to describe the electrical opposite of an electron. While an electron has a negative electrical charge, the positron has a positive charge. The electrical charge of a positron is equal to that of an electron only in terms of magnitude value (so called "absolute" value), but is opposite in sign. However the rest mass of the electron is equal to that of the positron in terms of both magnitude and sign. We should point out the difference between the positron, which has a positive rest mass, and the antielectron, which has a negative rest mass. By *antielectron* we mean the energetical and electrical opposite of the electron.

Under certain conditions, the pure energy of two weightless photons may be converted into a substantial pair consisting of a single weighty electron and a single weighty positron. Even when a substantial electron encounters a substantial positron, the two will be converted into a pure energy (non-substantial) pair of photons. The results of experiments of this sort have provided best confirmation of the law of transformation of weighty substance into weightless energy and weightless energy into weighty substance. They have once and for all put an end to the atheist myth of the imaginary "inconvertibility" of substance and energy.

On the other hand, if an electron encounters an antielectron, then both of them will be mutually and completely annihilated (they will disappear) without being transformed into anything material. In this process pure energy as well as weighty substance disappears (is annihilated). In this case *annihilation* means the complete mutual destruction (disappearance) of positive and negative opposites.

Energo-antiparticles such as antiphotons, antielectrons, and energo-antiprotons, were not and could not be experimentally observed, because of their very nature. If we tried to do so, our attempt would result in the total annihilation (mutual destruction) of the negative energy of the antiphotons (antielectrons or energo-antiprotons) and the positive energy of the substantial instruments. The energo-antiparticles and the instruments would disappear without being transformed into anything material. But this does not in any way mean that they do not exist. Elementary energo-antiparticles exist objectively, independently, and outside of any of our physical capabilities.

Once again this fact provides evidence of the truth that not every physical reality can be *detected by human instruments.*

To summarize the above, we can provide a concise statement of the *law of parity of elementary particles* which would go something like this:

1. *Every elementary particle has its own electrical, energetical, or other physical opposite, which we call an elementary antiparticle.*

2. *Elementary particles and antiparticles always appear and disappear in pairs: i.e., electron and positron, neutron and antineutron, photon and antiphoton, and so forth. No one particle can exist without an opposite.*

3. *The objective existence of elementary antiparticles has been confirmed experimentally. But the very nature of elementary antiparticles formed from negative energy renders them undetectable by material instruments with a positive rest mass. This example provides a graphic confirmation of the truth that not every (even physical) reality can be detected by human instruments.*

The occurrence and annihilation of particle and electric antiparticle pairs have provided splendid experimental confirmation for the law of the convertibility of matter and energy:

*Under certain conditions, any weighty physical body may lose its rest mass and be transformed into pure weightless energy. And conversely, weightless energy may acquire rest mass and be completely transformed into a weighty physical particle and then into a weighty physical body.*

So what does atheism have to counter the advances of modern science except for its own whims and illusions? To counter the scientific truth, it can only answer with a word game, which goes something like this: "If antimatter even exists, then it is only a particular type of matter, and not its opposite."

These sorts of word games are not science, but mere gibberish. Nevertheless, atheism has employed them successfully to mislead hundreds of millions of naive and gullible people.

If energo-antisubstance and negative energy are mere varieties of matter, then how can we call them antimatter (i.e., the opposite of matter)? If we still were to call them antimatter (i.e., the opposite of matter and not matter itself), then how could we believe the atheistic dogma, "There is nothing in the world except for matter in motion?"[2]

But names and gibberish cannot alter the truth. "Dialectical" materialism has solemnly proclaimed that nothing in the world can exist and evolve without its opposite. If materialism denies the existence of the opposite of positive energy, by doing so it ceases to be dialectical, and "scientific" atheism ceases to be scientific. In order to remain within the bounds of dialectics and science, materialism and atheism must recognize the objective existence of negative energy, no matter what they might call it.

# =4=

## THE FUNDAMENTAL LAW OF NATURE

*We used to call mass matter. But now it turns out
that this mass doesn't exist. Matter no longer exists.*

Henri Poincaré

I t would be impossible to prove or refute the initial premise of materialist philosophy scientifically without a clear definition of the concepts under comparison, namely "matter" and "idea." Therefore the first questions we might ask are: "What is the principal difference between matter and ideas? What are their most important attributes?" This chapter is devoted to a definition of the concept of matter, while Chapter 6 is devoted to a definition of the concept of idea.

In the face of new scientific discoveries, atheism broadened its definition of matter and included ordinary substance, antisubstance, waves, and fields in it. However, this expanded definition did not rescue atheism from scientific disaster. While physicists at the beginning of the twentieth century knew of only three building-block particles (the electron, proton, and photon), now we know of hundreds of them, and physicists are discovering new particles all the time. The atheists' revision of their concept of matter has literally not been able to keep up with the rapid pace of scientific discovery. Hence when all was said and done, atheism opted for a "wise" decision of calling any objective reality "matter," regardless of its nature.

We might ask now what objective reality is? If we use the word *triviality* to mean nonexistence as opposed to existence, then by *reality* we mean existence as opposed to nonexistence. Objective reality means literally everything which exists objectively, in reality, and factually. Hence "objective reality" encompasses not only animate and inanimate material systems but also any subjective or objective idea, including God. That is the reason why the "wisdom" of the modern atheistic definition of matter is truly infinite and knows no bounds. It encompasses everything that exists now and has ever existed. Even if scientists were to "discover" God, then this discovery would not pose the slightest threat to the atheistic dogma of the primacy of matter. In this extreme case atheism would have to acknowledge God merely as an objective reality, i.e., as a mere constituent of matter. But this discovery would not negate the primacy of matter. Modern atheism has tried to redeem itself from a scientific calamity by representing the idea as a variation of or even an attribute of matter, even though one of atheism's leaders asserted the opposite in his time.

Vladimir Ilyich Lenin wrote the following: "Calling an idea material means taking a step in the wrong direction of confusing materialism and idealism."[1] I might add the following: using the word "matter" to refer to the contents of any law of nature, which exists objectively outside of and independently of any subjective conscious-ness, means taking a step in the wrong direction of confusing gibberish and science. Calling an objective idea or God matter is the same as confusing religion and atheism. It is absolutely impossible to incorpo-rate the concept of idea in the concept of matter, which means that *matter is not and cannot be the only form of existence of objective reality.*

Nevertheless, by incorporating the concept of "idea" in the concept of "matter," atheism has consciously or unconsciously deprived us of the chance to compare these concepts. And by doing so, it has deprived us of any scientific opportunity to arrive at an objective solution to the very same problem which materialism considers fundamental, namely "What comes first: matter or idea?"

Indeed, we could definitely establish which of two hypothetical Smith brothers was born first, Bill or Tom. But it would be impossible to arrive at an objective solution to a problem such as the following: who was born first, Smith or Tom? This problem cannot be solved because Tom is the first name and Smith is the last name of the same person and thus he could not be born before or after himself. But if nevertheless a certain party states the problem this way and goes on to use this unclear statement of the problem as the basis for asserting that Smith (and thus Tom) was born before Bill, then this provides unmistakable evidence that the aforementioned "certain party" is not interested in the objective truth and wants to pass off his whims as the truth in the same "muddy water" which he himself has dirtied.

By perfect analogy, the atheistic definition of matter makes it impossible to arrive at a scientific solution of a basic philosophical question, namely "What comes first? Matter or idea?" It rules out any possibility of resolving this fundamental question of philosophy in idealism's favor and knowingly, without the slightest proof, includes a pat answer in favor of materialism. Consequently, it closes off scientific discussion of the issue. The irony of the materialist trickery lies in the fact that atheism has formally declared itself scientific while in reality its basic concepts and initial assumptions in effect rule out the possibility of any scientific inquiry. This makes it easy for atheism to pass off its whims as reality and its illusions as the truth.

However, the goal is the truth, regardless of what certain individuals might want. Scientific inquiry pursues no other goal. Hence we cannot emulate atheism's example and adopt its vague and slippery concepts of matter and idea following the atheistic principle of "matter is not a straightjacket; it will go wherever you want it to." If my fellow scientists and I seek objective truth scientifically, then we must define matter in a way which will allow us to distinguish matter clearly and unambiguously from the concept of idea.

Obviously, matter and idea both contain a multitude of distinctive attributes. It is also obvious that identifying the most essential attributes would pose no particular difficulty. But we are faced with the very difficult task of finding a correctly defined essential attribute of matter which would open up extensive opportunities for scientific inquiry into the truth and at the same time would be in congruence with some fundamental principles of atheism without which atheism would not be atheism. If we succeeded in giving matter this kind of scientific definition, then atheism would be forced to acknowledge it. If it acknowledged it and conscientiously followed the laws of logic, then atheism would inevitably proceed to scientific religion. But if atheism refused to acknowledge it, then it would have to reject its basic dogma. Thus, it would have to reject itself and depart from the world arena. Scientific religion would triumph either way. Therefore we face the difficult task of *giving matter a proper scientific definition which would prevent the shifty and crafty atheists from slithering away from or evading the objective truth.*

So, what is matter? What are the essential attributes which distinguish it from the idea? We will consider a few very simple examples in order to answer these questions.

**First example:**

Let us assume that yesterday I didn't have a penny to my name. Today I borrowed one hundred dollars from a friend. Now I have one hundred dollars in cash and one hundred dollars of debt. So, how much money do I have right now?

Merchants would answer this question by saying that I have a hundred dollars cash. These people are *relatively (but not absolutely!)* right because my debt has nothing to do with them. I could pay my cash and buy a hundred dollars worth of potatoes, onions, bread, and other goods from them. They are only right in the sense that this money is an objective reality and can be verified empirically, but I can dispose of it as I please.

Creditors would say that I am a debtor. They are also *relatively (but not absolutely!)* right because the debt is also an objective reality and can be verified empirically. At some point in time the creditor will demand repayment of the debt.

So both the merchants and the creditors are right. But both rights are only *relative but not absolute!* Each is right from his one's (relative) point of view, because both cash and debt are objective realities which can be verified empirically. Hence the amount of money I actually own, which in absolute terms is equal to zero, is trivial. The cash (+100 dollars) and the debt (-100 dollars) add up to zero. In other words, I still don't have a penny to my name. But what changed after I borrowed the money? What happened was that the zero assumed a completely new character and new quality. It is as if this trivial zero has "divided" into two real opposites. While before my net worth was equal to zero, now it has become a zero sum of material opposites. This means that a zero sum of opposites is not zero. They are fundamentally different from one another. A zero is a triviality which does not contain any components whatsoever, while a zero sum is an objective reality which consists of components which actually exist and are equal in magnitude but opposite in sign. In this case the components are cash and debt, which are fundamentally opposite to one another and at the same time could not exist without the other.

**Second example:**

An electron has a negative electrical charge, while a proton is positively charged. We might ask what sort of electrical charge a hydrogen atom, which consists of one electron and one proton, has.

Both the electron's negative charge and proton's positive charge are objective realities which can be verified experimentally. The electron's electrical charge exists objectively, in reality, outside of and independently of any subjective consciousness. But it exists only *in relation* to the proton's positive charge, and not in any *absolute* meaning of the word, because if there were no positive electrical charge, there would not be any negative electrical charge either. The proton's positive charge also exists objectively, in reality, outside of and independently of any subjective consciousness. But it also exists only *in relation* to the electron's negative charge, and not in any *absolute* meaning of the word, because if there were no negative electrical charge, there would not be any positive electrical charge either. The absolute existence of negative charges only or positive charges only (independent of their equivalent opposites) would be absolutely impossible! Therefore a hydrogen atom as a whole has no electrical charge. This means that the charge of a hydrogen atom is a zero sum of real opposites (positive and negative charges), which are fundamentally different, negate and at the same time affirm one another, and cannot exist without each other. For all the positive charges in the world, there are an equal number of negative charges, so that their sum is always equal to zero.

**Third example:**

Let us imagine some trivial point which has no energy whatsoever. Then let us assume that one quantum of positive energy and one quantum of negative energy are generated at this point, transforming the trivial point into an energetically closed system. One might ask what energy the system possesses.

Both negative and positive energy are objective realities because they exist objectively, in reality, outside of and independently of any subjective consciousness. But the existence of positive energy is *relative (and not absolute)*, because it presupposes the necessary presence of an equivalent amount of negative energy. The existence of negative energy is also *relative (and not absolute)*, because it presupposes the necessary presence of an equivalent amount of positive energy. The absolute existence of positive or negative energy only (independent of their equivalent opposites) would be absolutely impossible. The sum total of energy in the closed Universe has not changed in the least, has remained constant, and has always been and still is equal to zero. But what in fact has changed? The situation has changed only in the *qualitative* sense of the word and not the *quantitative* sense. While before the zero did not contain any energy constituents whatsoever, now it has apparently "divided" into a zero sum of energy opposites (negative and positive), which are radically different from one another, negate and at the same time affirm one another, and cannot exist without the other.

Thus, a trivial nothing (trivial zero) can be "split out" into a zero sum of any quantity of energetic opposites. The resultant system will not contain anything except for quanta of positive and negative energy. The rest mass of this quantum is equal to zero. However, according to the theory of Paul Dirac, the prominent physicist, under certain conditions this system may give rise to pairs of elementary particles and antiparticles with paired identical rest masses. For example, an electron is generated, exists, and disappears together with its electrical opposite, the *positron*. The electrical charges of the electron and positron are equal in magnitude but opposite in sign. But their positive rest masses are equal in both magnitude and sign. Thus, a purely energetic system can be "divided" into a zero sum of electrical charges, thus generating a system of elementary particles with a positive rest mass.

We have no idea of the total reserves of physical energy in the world. And it would be impossible for us to know this, and not simply because of our limited subjective (human) capabilities but for objective reasons which dictate that the amount of each type of energy (positive or negative) must be in a state of constant change. But if we examine the entire world as an isolated system, then we discover that the constantly changing amount of positive energy at any given moment in time is equal to the same constantly changing amount of negative energy so that their algebraic sum always remains constant and equal to zero.

**Fourth example:**

From physics we know that every elementary particle has its own equivalent opposite, and that elementary particles occur and disappear only in pairs with corresponding antiparticles, i.e. electron and positron, proton and antiproton, neutrino and antineutrino, and so forth. No particle can exist without its opposite. This means that any physical substance is either explicitly or implicitly a zero sum of such real opposites which are equal to one another in magnitude but opposite in sign.

I would therefore ask what distinguishes the existence of the physical world (as the aggregate of all elementary particles and antiparticles) from its nonexistence?

Any pair consisting of a particle and an antiparticle contains a zero sum of real opposites which are equal in magnitude (in quantity) but opposite in sign (in quality). Any substance and the entire physical world are constructed from a multitude of these elementary particles and antiparticles. We also know that the sum of any infinitely large number of zeros is always zero. Hence the entire physical world (no matter how big it may be!) is a zero (trivial) sum of real (non-zero) opposites.

Thus, if the nonexistence of the physical world can be represented as the complete absence of all particles and all antiparticles, all substance, and all antisubstance, then the existence of the physical world constitutes a zero sum of non-zero (actually existing!) opposites such as substance and antisubstance, particles and antiparticles, positive energy and negative energy, and so forth.

**Fifth example:**

Human sexuality is an objective reality that can be verified empirically. Sexuality, however, exists only in *the relative but not the absolute* sense of the word, because the sexual activity of the male only exists in relation to the female, and the sexual activity of the female only exists in relation to the male. To an outside observer, the sexual activity of a couple would be equal to zero. This means that the sexual activity of the couple is a zero sum of real opposites.

In the absolute sense of the word, the sum of the sexual activity of the couple is equal to zero before *and* after the union. But what in fact has changed during such union? Before the union, the sexual activity of both was equal to zero. Afterwards, the situation changed in qualitative terms. Now we have two sexual opposites. Even though the sum is still equal to zero, this zero sum is not zero and is not trivial because it consists of real sexual constituents which are equal in magnitude and opposite in sign.

While men and women have fundamentally different sexual attributes, they always affirm one another and could not exist without the other. While man's sexual activity exists objectively, in reality it exists only *in relation to the woman's, and not in the absolute sense of the word.* If there weren't any women, there would not be any men, and vice versa. The absolute existence of only men or only women (independent of their equivalent opposites) would be altogether impossible! The sexual capacity of all men and women as a whole is a zero sum of such real opposites, which are radically different from one another, negate and at the same time affirm one another, and cannot exist without the other.

An electrical model of male and female sexuality might be constructed of two physical items, where the first gradually accumulates a positive and the second a negative charge. The contact of the two charges could serve as an electrical model of the sex act. We could arrive at similar conclusions after examining any other aspect of the world around us. Then, by using the scientific method of induction, viz., deriving a general rule from particular examples, we would define the *fundamental law of nature* in the following way: *There is no physical category which can exist or develop without its opposite.*

The mandatory presence of opposites in explicit or implicit form is not the source or driving force of evolution but is merely a necessary (but not sufficient!) condition for the existence and development of any physical reality.

*Matter is the zero sum of real opposites which are radically different from one another in terms of some attribute, negate and at the same time affirm one another, and cannot exist without the other. Any form of existence of matter is relative. An absolute form of the existence of matter is an impossible category.*

We will therefore use the term *matter* to refer not to any objective reality, but only to the kind which *cannot* exist and develop without its opposite.[2] An electron could not exist without a positron; a negative electrical charge could not exist without a positive charge; positive physical energy could not exist without negative energy, and so forth.

We refer to energy as matter, even though it doesn't have any volume or weight. Weightless energy existed in the newborn Universe without its opposite. But as the positive weightless energy develops, it is bound to evolve into its dialectical opposite, i.e., weighty substance. This is the only thing it can do. And, again, energy is a material category.

Life forms which absorb oxygen and release carbon dioxide could not exist without green plants, which absorb carbon dioxide and release oxygen. On the other hand, green plants could not exist without these carbon dioxide "producers." Man cannot exist without woman, the seller cannot exist without the buyer, without students there are no teachers, and so on.

Henri Poincaré, the prominent French physicist and mathematician, has written the following on the subject: "We used to call all mass matter. But now it turns out that this mass doesn't exist. Matter no longer exists."[3] Max Ernst, the prominent Austrian physicist, believed that all "neutral elements" are nothing.

"Dialectical" materialism, and thus atheism, have acknowledged this indispensable attribute of matter. Moreover, they consider opposites to be an essential source for any dialectical development. Therefore, if materialism were to disagree with the definition of matter given above, then it would inevitably cease to be dialectical, and atheism would inevitably cease to be scientific. But if atheism were to acknowledge the basic attribute of matter cited above, then it would inevitably be compelled to agree with the scientific conclusions of religion. The objective truth would triumph in either case. Atheists themselves have written the following concerning the basic property of matter:

> In dialectics the term 'opposites' means the aspects, characteristics, or features of any object which basically differ in terms of a specific attribute and at the same time affirm one another and cannot exist without the other. Every thing, every phenomenon contains opposites within itself. In mathematics, for example, there is the positive and the negative, the differential and the integral, while in mechanics there is the action and reaction, in physics there are positive and negative charges, in chemistry there is the combination and breakdown of molecules, and so forth. Opposites are universal in nature and are inherent in all objects and phenomena of reality.[4]

This quote makes it absolutely clear that our definition of matter is completely consistent with the basic principles of the dialectics upon which both modern materialism and its progeny, i.e. scientific atheism, are so completely reliant. The only essential difference between our definition of matter and the atheist definition is that ours does not try to encompass all of objective reality *a priori*, which means that it does not include any cut-and-dried answer in favor of idealism or materialism and thus opens up extensive opportunities for a scientific search of objective truth. Obviously, the necessary presence of opposites is not the only important property of matter, which undoubtedly has very many other significant attributes. For example, all of the material elements and systems that we know contain energy. Hence matter could be called a reality which possesses energy, mass, and so forth.

Modern 'scientific" atheism, however, which has succeeded in deceiving half the population of the globe, does not rely on such plain and simple concepts as weight and volume, which are absolutely irrelevant to atheism's existence, or moreover, its prosperity. The philosophical foundation of modern atheism is "dialectical" material-ism. This means that "scientific" atheism is perched on a firm branch whose name is dialectics. All nondialectical forms of materialism have been discredited to such an extent as to make the existence of scientific atheism without dialectics absolutely unthinkable. In turn, dialectics sits on another sturdy branch and this branch is called "the mandatory presence of opposites." Atheism could never abandon a property of matter such as the necessary presence of opposites. If it did, it would cut off the branch on which it is so comfortably perched.

And if in spite of all this, the principle of "opposites" and the dialectics based on this principle prove to be in glaring conflict with materialism, then atheism would have nowhere to run. Its only way out would be to declare bankruptcy. That is why, of all the most important properties of matter we have singled out, the one most important property in philosophical aspects is the property of opposites.

We should also point out that the definition of the fundamental property of matter is not an unsubstantiated assertion or an unproven axiom. It is the result of conscientious scientific generalization by means of the generally accepted inductive method.

With what can "scientific" atheism counter the fundamental law of nature? With nothing but contradictory quotations from its leaders. However, since Lenin himself, the now-discredited leader of Russian atheism, claimed that "to refer to thought as material means making an erroneous step toward confusion between materialism and idealism,"[5] how can one believe him when he says that "there is nothing in the world except for matter in motion?!"[6]

# THE LAWS OF CONSERVATION AND CREATABILITY OF MATTER

*The highest duty of physicists is to search for those elementary laws from which one can obtain a picture of the world by means of pure deduction.*
Albert Einstein

S o-called scientific atheism has constructed its "theories" on the initial premise of the "uncreatability and non-annihilability of matter." This initial premise has never been proven theoretically or confirmed experimentally by anyone, which means that it is not the result of any scientific proof whatsoever. Atheists and materialists, however, have sometimes referred to the laws of conservation of matter or energy in order to impart some sort of scientific aura to their fabrication. Thus, they have tried to substitute the far from equivalent term "uncreatability of matter" for the term "law of conservation of matter" in the minds of naive and gullible individuals. Therefore it would be quite appropriate to ask whether the scientific laws of conservation have in fact confirmed or discredited the materialist concept of the "uncreatability of matter."

We know of a large number of laws of conservation from the natural sciences. But all of these laws follow the same general principle. We will discuss a few of them as examples.

**Examples from classical mechanics:**

1. From the law of conservation of impulse of force it follows that the vector sum of all forces acting on the elements of an isolated system does not change in time. This means that in an isolated system

it would be impossible to increase (or decrease) any force without causing a simultaneous and equivalent decrease (or increase) in another force. But this law provides no evidence of "uncreatability" or "non-annihilability" of force at all. On the contrary, it permits an increase (or decrease) in any force as long as it is accompanied by a simultaneous and equivalent increase (or decrease) in its opposite provided that the vector sum of the two in a multidimensional space remains constant. In the simplest case, when all of the forces act along the same line, their vector (geometric) sum is equal to the algebraic sum. There is no sense in considering extremely complicated systems in this popular and commonly accessible discourse of ours. Therefore, the concept of "algebraic sum" will henceforth refer to the simplest schemes, unless specifically indicated otherwise.

For example, the algebraic sum of all forces operating on a cannon and a projectile along a certain axis **x** will remain constant and equal to zero both before and during the firing of the projectile. While there were no forces whatsoever before the projectile was fired ($F_1 = F_2 = 0$), the firing of the projectile resulted in the appearance of a zero sum of opposite forces, i.e., $F_1 + F_2 = 0$ or $F_1 = -F_2 \neq 0$. One force operates on the projectile, while the other (opposite) force operates on the cannon.

Here is yet another example. Suppose, an electron and a positron move by inertia to encounter each other. Force, applied to each of them, is equal to zero. Thus, the algebraic sum of both of them is also equal to zero. The collision of an electron and a positron transforms them into two photons, which move in different directions at the speed of light. At the first moment of rebound the impulses of these photons are equal in magnitude and opposite in sign. The pair of opposite impulses comes into existence literally from nothing, even though their algebraic sum remains constant and equal to zero.[1]

This means that *in an isolated system impulses and proportional physical forces may appear (or disappear) as long as their opposites appear (or disappear) simultaneously and proportionally so that their algebraic sum always remains constant.*

2. The law of conservation of momentum means that the algebraic sum of the momentum of an isolated system is always constant. This means that it would be impossible to increase (or decrease) the momentum of any element of an isolated system without causing a simultaneous and equivalent decrease (or increase) in the momentum of another element.

But this law provides absolutely no evidence whatsoever of the "impossibility" of the appearance or disappearance of momentum. On the contrary, according to this law, the momentum of any element of a conservative system may appear (or disappear) as long as it is accompanied by the simultaneous and equivalent appearance (disappearance) of its opposite, so that their algebraic sum always remains the same. For example, the algebraic sum of the momenta of a loose cannon and projectile will remain constant and equal to zero both before and after the projectile is fired. While there was absolutely no motion before the projectile was fired ($m_1v_1 = m_2v_2 = 0$), then the firing of the projectile resulted in the appearance of a zero sum of opposite momenta: $m_1v_1 + m_2v_2 = 0$ or $m_1v_1 = -m_2v_2 \neq 0$.

3. The laws of the conservation and appearance (or disappearance) of angular momentum can be stated similarly:

The angular momentum of an isolated system is always constant.

The angular momentum of any element of an isolated system may appear (or disappear) as long as it is accompanied by the simultaneous and equivalent appearance (or disappearance) of its opposite such that their algebraic sum is always constant.

**Examples from physics:**

4. According to the laws of conservation, no elementary particle

may be generated (or annihilated) without the simultaneous genera-tion (or annihilation) of its opposite, i.e., without its corresponding antiparticle. But this in no way implies that the number of pairs of elementary particles and antiparticles will remain constant and unchanged now and forevermore. On the contrary, the law of conservation is opposed by another (opposite) law of nature, namely the law of the generation and disappearance (annihilation) of elementary particles, which stipulates that the number of elementary particles and antiparticles is constantly changing so that the number of generated particles will always remain equal to the number of generated antiparticles and the number of disappearing (annihilated) antiparticles will always remain equal to the number of annihilated (disappearing) particles. The laws of conservation and creation are opposite categories of the fundamental essence of matter which could not exist without the other.

The possibility of the generation or annihilation of each pair of opposites (particles and antiparticles) is protected by a certain series of conservation laws. For example, the possibility of the generation or annihilation of a positron and electron is protected by the following five laws of conservation: energy, impulse, electrical charge, lepton charge, and spin. Let us discuss several of these laws as examples.

5. The law of conservation of electrical charges states the follow-ing: *in an electrically isolated system the algebraic sum of electrical charges will always remain constant.* This law says that in an electrically isolated system it would be impossible to increase (or decrease) the number of negative electrical charges without simultaneously and proportionally increasing (or decreasing) the number of positive electrical charges. And conversely, it would be impossible to increase (or decrease) the number of positive electrical charges without simultaneously and proportionally increasing (or decreasing) the number of negative electrical charges. This law provides absolutely no evidence whatsoever of any sort of "uncreatability" or "non-annihilability." On the contrary,

it allows any increase (or decrease) in one opposite as long as it is accompanied by a simultaneous and equivalent increase (or decrease) in the other opposite such that their algebraic sum always remains constant.

For example, from physics we know that the annihilation (disappearance) of every electron-positron pair causes an electrically isolated system to lose one positive and one negative electrical charge. *The pair of opposite electrical charges literally disappears (annihilates) and is not transformed into anything else.* The opposite process, i.e. the generation of each electron-positron pair, causes an electrically isolated system to acquire one positive and one negative charge. *This pair of opposite electrical charges literally appears (comes into existence) from nothing.*

According to the laws of dialectics, the numbers of each type of electrical charge (positive or negative) should constantly change while their algebraic sum will remain the same. This means that the law of conservation of electrical charges has an opposite known as the law of creatability and destructibility of electrical charges which states that *any type of electrical charge is creatable (annihilable) as long as it is accompanied by the simultaneous and equivalent creation (annihilation) of its opposite.*

Electrical charges have been seen as matter even by materialists like Lenin (who stated that there is nothing except matter in motion), while fundamental physics has experimentally proven that a pair of opposite electrical charges can be generated out of nothing or disappear without being transformed into anything. Consequently, *matter is creatable and annihilable.*

We do not know how many positive or negative electrical charges there are in the world. And it would be impossible for us to know, not just because of the limitations of our subjective capacities but for such objective reasons as the fact that the number of each kind of electrical charge must always be in a state of constant change. However we do know that *particles and antiparticles are "born" and "die" only in pairs.*

This means that the numbers of electrons and positrons in the world must be identical. If we consider the entire Material World as an electrically isolated system, we know that the number of negative electrical charges in it is constantly changing but at any given moment in time is equal to the constantly changing number of positive electrical charges so that their algebraic sum is always constant and equal to zero.

6. The same applies to the law of conservation of baryon charges. *A pair of opposite baryon charges literally annihilates without being transformed into anything else.* The opposite process, i.e. the generation of each proton and antiproton pair, causes an isolated system to acquire one positive and one negative baryon charge. *A pair of opposite baryon charges literally appears (comes into existence) from nothing.*[2] According to the laws of dialectics, the number of each kind of baryon charge (positive or negative) should constantly change while their algebraic sum should remain constant. This means that the law of conservation of baryon charges has its opposite known as the law of creatability and destructibility of baryon charges, which stipulates that *any kind of baryon charge can be created (or annihilated) as long as this creation (or annihilation) is accompanied by the simultaneous and equivalent creation (or annihilation) of its opposite.*

7. The same applies to the law of conservation of lepton charges. *A pair of opposite lepton charges literally appears (comes into existence) from nothing.*[3] According to the laws of dialectics, the number of each kind of lepton charge (positive or negative) should constantly change while their algebraic sum should remain constant. This means that the law of conservation of lepton charges has its opposite known as the law of creatability and annihilability of lepton charges, which states that *any kind of lepton charge can be created (or annihilated) as long as this creation (or annihilation) is accompanied by the simultaneous and equivalent creation (or annihilation) of its opposite.*

8. The law of conservation of spin states the following: *in a*

*conservative system the algebraic sum of all half-spins is always constant.*
*Spin* refers to the intrinsic angular momentum which elementary
particles possess. Therefore, this law is essentially a special case of the
law of conservation of angular momentum. This law says that in a
conservative system it is impossible to increase (or decrease) the
number of negative half-spins without simultaneously and proportionally increasing (or decreasing) the number of positive half-spins. And
conversely, it would be impossible to increase (or decrease) the number
of positive half-spins without simultaneously and proportionally
increasing (decreasing) the number of negative half-spins. But this law
provides absolutely no evidence whatsoever of the supposed "uncreatability"or "non-annihilability" of spins! On the contrary, it allows for
any increase (or decrease) in one opposite as long as it is accompanied
by a simultaneous and equivalent increase (or decrease) in the other
opposite such that their algebraic sum always remains constant.

For example, from physics we know that a negative muon decays
into an electron, neutrino, and antineutrino.[4]   The annihilation
(disappearance) of every neutrino and antineutrino pair causes a
conservative system to lose one positive and one negative half-spin.

*A pair of opposite half-spins literally disappears* (annihilates, vanishes,
passes out of existence) *without being transformed into anything else.* The
opposite process, i.e. the generation of each neutrino and antineutrino
pair, causes a conservative system to acquire one positive and one
negative half-spin. *A pair of opposite half-spins literally appears* (comes
into existence) *from nothing.*[5]  According to the laws of dialectics, the
number of each kind of half-spin (positive or negative) should
constantly change while their algebraic sum should remain constant.
This means that the law of conservation of half-spins has its opposite
known as the law of creatability and destructibility of half-spins, which
states that *any kind of half-spin can be created (or annihilated) as long as
this creation (or annihilation) is accompanied by the simultaneous and
equivalent creation (or annihilation) of its opposite.* Once again, *matter is*

*creatable and annihilable.*

All of the other laws of conservation from physics or the other natural sciences which we might examine would inevitably lead to similar conclusions.

## Examples from the social sciences

9. In studying the development of human society, we could formulate the following law of conservation of society's sexual potential: *the algebraic sum of the sexual potentials of human society as a whole always remains constant.*

This means that in human society the number of males could not increase (or decrease) without a proportional increase (or decrease) in the number of females. And the number of females could not increase (or decrease) without a proportional increase (or decrease) in the number of males. Under normal conditions more boys are born than girls, which means that the death rate for boys is higher than the death rate for girls. In wartime, when large numbers of males are killed, the percentage of male births would increase and the percentage of female births would decrease in order to restore the artificially upset balance of sexual opposites.

According to the laws of dialectical development, the number of all males and females must constantly change while the "algebraic sum" of sexual potentials must remain constant. This means that the law of, conservation of sexual potentials has its opposite, namely the law of creatability of sexual potentials, which stipulates that *in human society the number of males can increase (or diminish) only in proportion to the increase (or decrease) in the number of females. And conversely, the number of females can increase (or diminish) only in proportion to the increase (or decrease) in the number of males.*

We could cite an infinite number of examples of this type. No matter how many examples we study, we would always arrive at the same conclusions. So, by using the generally accepted scientific method of induction, we may proceed from particular examples to the following *generalized laws of conservation and creatability of matter.*

**The law of conservation of matter:**
*In an isolated system, the algebraic sum of the quantity of all forms of matter is always constant and invariable.*

**The law of creatability of matter:**
*Any kind of matter can be created (or annihilated) as long as its creation (or annihilation) is accompanied by the simultaneous creation (or annihilation) of an equivalent amount of its material opposite.*

The laws of conservation and creation of matter are opposite categories of the same essence of indivisible matter which could not exist without each other. Therefore, the law of creatability of matter is fully consistent with the law of conservation of matter. The fallacious assumption of the "uncreatability" of matter is completely discredited by the law of conservation of matter as well as the law of creatability of matter.

According to the laws of conservation and creatability of matter, regardless of how or to what extent matter may have developed, the algebraic sum of all material opposites in the world is always trivial and always equal to ideal zero.

The following fundamental property of matter proven in the previous paragraph, follows directly from the generalized laws of conservation and creatability: *no material category can exist or develop without its internal and external opposite.*

We will therefore use the term *matter* to refer only to that objective reality which cannot exist and develop without its internal and external opposite.

Albert Einstein wrote that "The highest duty of physics is to search for those elementary laws from which one can obtain a picture of the world by means of pure deduction."[6] Above, we formulated precisely these kinds of generalized laws of conservation and creation of matter, from which, by means of pure deduction, we can now derive both the fundamental law of nature, and the "world pattern," including laws of the conservation and creation of energy.

The law of conservation of energy states the following: *in an energetically isolated system, the algebraic sum of all forms of energy is always constant.* This law says that in an energetically isolated system positive energy cannot be increased (or decreased) without a simultaneous and equivalent increase (or decrease) in negative energy. Conversely, negative energy cannot be increased or decreased without a simultaneous and equivalent change in positive energy. No kind of positive energy (such as thermal energy) can be increased (or decreased) without a simultaneous and equivalent decrease (or increase) in another form of positive energy (such as mechanical), and so forth.

The scientific law of the conservation of energy, however, provides no corroboration whatsoever for the erroneous belief in the "uncreatability" or "non-annihilability" of energy. On the contrary, it allows for any increase (or decrease) in positive energy as long as it is accompanied by a simultaneous and equivalent increase (or decrease) in negative energy, even though their algebraic sum always remains the same. At the same time, according to the laws of dialectics, the amount of each type of energy (positive and negative) must be in a state of constant change. This means that the law of conservation of energy must have its opposite, namely the law of the creatability and destructibility of energy, which states that: *positive energy can be created (or annihilated) as long as this creation (or annihilation) is accompanied by the simultaneous and equivalent creation (or annihilation) of negative energy. And conversely, negative energy can be created (or annihilated) as long as this creation (or annihilation) is accompanied by the simultaneous and equivalent creation (annihilation) of positive energy.*

We have no idea of the total supplies of energy in the entire

Universe. And it would be impossible for us to know, not just because of the limitations of our subjective capacities, but for such objective reasons as the fact that the amount of each type of energy (positive or negative) must be in a state of constant change. But if we consider the material world as a whole as an isolated system, then we would know that the continuously changing amount of positive energy is, at any given moment in time, equal to the continuously changing amount of negative energy such that their algebraic sum is always constant and equal to zero.

If all of the weighty substantive particles of the world were to encounter their electrical antiparticles, all of them would be trans- formed into pure and weightless energy. Then this type of purely energetic world would be a zero sum of a set of weightless particles such as photons and antiphotons. If this kind of purely energetic world were compressed to hypercritical density into a single small "cosmic egg," as the atheists call it, then all the positive energy and all the negative energy in the entire world would completely annihilate each other without being transformed into anything material. Then our physical world would completely cease to exist.

"How strange! But it's even stranger that our world and we ourselves still exist after all this," wrote the Soviet scientist Vitaliy Isaakovich Rydnik.[7] And in reality you and I have never seen this desolate picture of the disappearance of the physical world.

But why?

Atheist scientists have responded by saying that "provident nature did everything it could to keep positrons and electrons," particles and antiparticles, substance and antisubstance, galaxies and antigalaxies as far apart as possible. If not, our physical world and we ourselves would cease to exist. But then I would ask how inanimate nature could have a mind so as to "provide" any sorts of safeguards to prevent a catastro- phe. Obviously only a mind can "provide." Inanimate nature has absolutely no intelligence and no mind whatsoever. Consequently, instead of the physical world itself or blind nature itself, some other nonphysical (nonmaterial) category with a higher intelligence must be saving the physical world from disaster. And we use the term *God* for this nonphysical category which is saving the world from such an incredible disaster, and we use the term *Divine Intelligence* for the great mind which has so brilliantly provided all the safeguards for the physical world.

If we were to arbitrarily use the term positive energy to refer to the energy of the sun, of machinery, and of all the objects around us, then we might justifiably ask: where is the negative energy?

We can get the answer to our question from the theory of Paul Dirac, the British physicist. According to his theory, the physical space of the Universe (which is incorrectly called vacuum space) is a continuum of dispersed negative energy which has no rest mass and no weight. The atheistic interpretation of the law of conservation of energy assumes implicitly and *a priori* that the world only contains positive energy and has absolutely no negative form of energy in vacuum space, which contradicts Dirac's theory and the facts of modern physics.

# =6=

## IDEA AND MATTER

*Everything that is not an idea is pure nothing...*
Henri Poincaré

A theistic materialism has constructed its antireligious theories on the unsubstantiated initial premise that "there is nothing in the world except for matter in motion."[1] But then it might be appropriate to ask: what does modern science say about the subject?

To modern scientists, the photon is a unique "microworld" with its own unique "microcivilization," even though its physical volume is equal to zero.[2] Weighty and visible substance is the crudest form of objective reality. Weightless energy is an objective reality of a higher, more refined quality than weighty and visible substance.

If we were able to split the energy "micro-world" of the photon into "component parts," ultimately we would discover an even more complicated and more perfect reality which in itself contains nothing material whatsoever (not even physical energy!). This nonmaterial category might be both the "objective idea" and the "ideal spirit." In the other world, the ideal spirit differs from the objective idea in approximately the same way as a living human being differs from inanimate matter in this world.

We will use the term *idea*, which may be both subjective and objective, to refer to any nonmaterial opposite of matter. We will call any idea which belongs to the human mind (or the mind of any other biological creature) *subjective*. An example of a subjective idea is the semantic content of an invention produced by an engineer.

*The fundamental law of nature has the following corollary. If matter "cannot" exist without its opposite, then there must exist some opposite of matter which "can" exist without its opposite. We will use the term objective idea to refer to this kind of nonmaterial opposite of matter which "can" (but does not necessarily) exist without its own relative opposite.*

An objective idea exists outside of any subjective (human) consciousness and is completely independent of it. A classic example of an objective idea is the semantic content of the laws of nature, including all those laws which have not yet been discovered by human (subjective) intelligence. The objective idea, which is completely independent of any observer, even existed when there were no humans and no Universe at all. The subjective idea is confined to the material brain, is a reflection of it, and constitutes an imprecise (or sometimes distorted!) copy of a precise objective idea.

A beam of light passing through glass is to some extent distorted by the glass in the process. The degree of distortion depends on the quality of the glass. But this does not in any way mean that the beam of light could not exist without the glass.

If a picture which passes through a TV screen is an imprecise copy of the actual outward appearance of a human being, the degree of distortion depends on the quality of the television set. But this does not in any way imply that the outward appearance of a human being could not exist without the TV.

By perfect analogy, if some ideal information acquired by a material brain is reflected by it with a certain amount of distortion, and the reflected information is a subjective (imprecise) copy of an objective (precise) idea, then the degree of distortion depends on the quality of the brain. But this does not by any stretch of the imagination mean that an objective idea perceived by a material brain supposedly cannot exist without a material brain. Only a subjective idea reflected by the material brain depends on the brain, cannot exist without it, and therefore possesses some of the properties of matter. At the same time a subjective idea, as a copy of an objective idea, possesses many of the qualities of an objective idea and can exist without the other forms of its opposite.

The quality of an objective idea may be conceived as the limit toward which the quality of a subjective idea tends as the material brain develops but which it will never reach.

While an atheist might protest against the concept of objective ideas and assert that any idea and any semantic content supposedly exists only in the human (subjective) brain, I will be so bold as to question him as to the whereabouts of the semantic content of the laws of nature, which have not yet been discovered by human intelligence but which gave birth not only to man himself but all life.

But even an objective idea which exists in relative space and in relative time is only a relative category and therefore cannot exist and evolve without its absolute opposite. The perfection of the *absolute idea* may be conceived as the limit toward which the perfection of an objective (relative) idea tends in its perpetual evolution but which it will never reach.

A subjective idea, like matter, must have a beginning and an end. An objective idea needs a creator but must not accumulate error. Therefore, an objective idea must have a beginning, but does not necessarily have an end. An absolute idea has neither a beginning nor an end.

We will use the term *reality* to refer to any matter and ideas. It is equally applicable to both, i.e. matter and idea are united by the common term *reality*.

We will use the term *physical Universe* or simply the *Universe* to refer to the totality of all those material particles, elements, systems, planets, stars, and galaxies without any exception which have the same common physical space.

According to the theory of N.S. Kardashev, a Corresponding Member of the Soviet Academy of Sciences,[3] the *Material World* consists of a finite set of physical Universes such as ours. Universes may differ from one another qualitatively, i.e., they may have different laws and different space-time relationships.

Unlike other Universes, we write our Universe starting with a capital letter.

According to the fundamental law of nature, our physical Universe could not exist and develop if it did not have a nonmaterial opposite known as the ideal Universe. Therefore, Universes may be both physical and ideal. We will use the term *ideal Universe* to refer to the total combination of all, without any exceptions, ideal elements and systems which exist objectively and have the same common autonomous nonphysical space.

We will use the term *Ideal World* to refer to the total combination of all ideal Universes, including all ideal systems and all ideal elements.

Thus, the Ideal World consists of a set of ideal Universes. One of these is the ideal Universe where our nonmaterial souls live and develop.[4] We will use the term *ideal space*, which in contrast to vacuum (physical) space, contains nothing material and nothing physical whatsoever, even energy, to refer to the domain of existence of all ideal categories.

If there were no idea and Ideal World, there would be no matter and Material World. According to the fundamental law of nature, the objective idea and Ideal World are indispensable for the existence of matter and the Material World. But from our daily experience we know for sure that matter and the Material World exist in reality and objectively outside of and independently of any subjective (human) consciousness. Consequently, the idea and the Ideal World exist just as objective and just as independent of our subjective desires.

If we acknowledge the objective existence of matter, the physical Universe, and the Material World, we must equally acknowledge the objective existence of their nonmaterial opposites such as idea, the ideal Universe, and the Ideal World. Either we acknowledge the scientifically proven fact of the actual existence of the objective idea, ideal Universes, and the Ideal World, or we must go against the scientific facts and reject science and the fundamental law of nature, whose truth is obvious to everyone. We have no other alternatives.

Usually people conceive of the Ideal World as an extension of the Material World in the same way that the territory of Belorussia begins once you get past Russia, the territory of Poland begins once you get past Belorussia, and so forth. But in reality this is far from the truth, because the Ideal and Material Worlds are *qualitatively different* and have fundamentally different dimensions. The physical space of the Material World, which is measured in kilometers or miles, is qualitatively different from the space of the Ideal World, which cannot be measured in kilometers or miles, in the same way that the mind cannot be measured in kilograms.[5]

An idea may be positive or negative. A positive idea exists independently of both matter and negative idea. It may exist not only in the complete absence of a negative idea, but also in its presence.

A negative idea (in contrast to the positive idea) cannot exist in the presence of its positive opposite. A negative idea may only exist when it has no positive opposite. If a positive idea increases as it develops, its negative opposite will decrease and asymptotically approach zero. For example, the truth exists independently from the lie, but a lie can exist only if it is not opposed by the truth.

In the principle of opposites materialism is right where, when, and to the extent that it acknowledges that, consistent with the laws of conservation of matter, *no material category can exist and develop without its opposite. Primordial matter could not be created from nothing as a zero sum of non-zero antipodes if it did not have a nonmaterial opposite, i.e., the objective (nonhuman!) idea.*

But when materialism automatically extends the principle of opposites to nonmaterial ideas, it is far from always right, because there is no such thing as the law of conservation of ideas in the world. Therefore an idea (in contrast to matter) can, but does not necessarily, consist of opposites. For example, the truth can, but does not necessarily, coexist with a lie. Human intelligence or the human mind exists without its ideal antipode, because there is no such a thing in the world as "anti-intelligence" or "anti-mind."

At the same time materialism is completely wrong where, when, and to the extent that it considers opposites to be the source of the development of matter or an idea. In fact, the primordial source of all development is the creative activity of an absolute God, and not the componential opposites of relative matter. The presence of opposites is not a source of development but merely a necessary condition for the creation and development of matter.

An essential difference between an ideal category and a material category is also found in the fact that a positive idea can exist without a negative opposite, while matter cannot. Hence matter is dependent on the idea, while a positive idea is independent of both matter and its negative opposite. Moreover, the presence or development of a positive idea discredits any negative idea, while any material category inevitably accumulates error, which results in the disintegration and disappearance of a material system.

According to the laws of conservation of matter, regardless of how and to what extent matter may develop, the algebraic sum of all material opposites in the world is always trivial and equal to ideal zero. According to the law of transience, matter cannot exist forever. At the same time, according to the fundamental law of nature, the zero sum of transient matter could not exist without its eternal and infinitely large opposite. But eternity and infinity are impossible categories for the Material World. Therefore we must look for them outside of any matter, outside of the entire Material World, in a qualitatively different Ideal World. From this we may arrive at the scientific conclusion that the objective idea is in a state of perpetual development and that the Ideal World is infinitely large.

As a result of the infinite development of objective ideas, the Ideal World has become infinitely large. While we can express the "algebraic sum" of all kinds of ideas in the world as an infinitely large quantity, the algebraic sum of all forms of matter in the world is equal to ideal zero. The Material World is trivial.

Transient matter is not eternal but rather trivial, and is dependent on the idea, while the idea is eternal, infinite, and independent of matter. Therefore any invisible, ideal category, which has no weight and no volume, is more perfect than any visible material category, which has weight and physical volume. All of this should unequivocally convince us that the idea and the Ideal World are objectively real to a much greater extent than matter and the Material World.

An objective idea (or "ideal spirit") does not have any material attributes. It cannot be seen with the eyes, heard with the ears, touched with the hands, or detected by instruments. Nevertheless any idea is a reality of a higher and more perfect quality than physical energy.

## Example

Modern biologists have definitely established that a living cell contains a system of genetic codes whose semantic content is known as the *genetic program*. This program completely determines the behavior and development of a living organism.

While a genetic code recorded at the level of weighty molecules or weightless photons is a material ("thisworldly") category, the genetic program (as the nonmaterial opposite of the material code) is an ideal ("otherworldly") category. Here, as in conversational speech, the term "idea" is used to express not just the nonmaterial opposite of the material code but its semantic contents as well.

The system of genetic codes is only the material record of an ideal program, while the genetic program is the semantic essence of this molecular record. An ideal program can exist without a material record: an example of this is provided by the large number of unwritten (ideal) laws of material nature. But a material record could not exist without a semantic content, because no record whatsoever could exist without semantic contents.

According to the fundamental law of nature, this system of genetic codes as a material category could not exist without an ideal opposite, which we call the *genetic program*.

The genetic program is an ideal category, while its code is a material category. While a genetic program (as an ideal category) is completely free of any error and is in a state of constant development, its genetic code (as a material category) cannot be freed from its destructive opposite, namely cumulative error. Hence a genetic program may be ideally perfect, but not its molecular (material!) record, i.e., the system of genetic codes. The perfection of the genetic program is that unattainable limit toward which the development of genetic codes constantly tends but will never reach.

While molecules and genetic codes, which have some sort of elementary volume, can only be seen under a microscope, the ideal contents of the genetic program itself, whose physical volume is equal to ideal zero, cannot be "seen" at all. We can only detect it mentally, on the basis of scientific conclusions, using our intelligence and our minds. Nevertheless an ideal genetic program is much more perfect than its material code, which is recorded at the level of weighty molecules or weightless photons.

On the basis of current data from the natural sciences, we may summarize the law of the objective existence of ideas and the Ideal World as follows:

1. *Any objective reality may be divided into two basic categories: idea and matter. Any existence may be material or ideal. Consequently, the entire Relative World consists of two basic opposites: the Ideal World and the Material World.*

2. *According to the fundamental law of nature, the actual existence of an objective idea and the Ideal World is not just a definite fact but an absolute necessity, without which the existence and development of*

*contradictory matter and the Material World would be impossible.*

3. While according to the laws of conservation of matter, the *Material World is only the zero sum of a countless set of physical opposites, the Ideal World is an infinitely large positive quantity as a result of its infinite development. Hence the idea and Ideal World are objective realities to a much greater extent than matter and the Material World.*

4. *While weightless and invisible physical energy may be detected by physical instruments, nonmaterial ideas cannot be detected even by instruments. Nevertheless, weightless and invisible ideas are objective realities of a more perfect and higher quality than weighty and visible matter, even though they cannot be directly seen by the eyes, heard with the ears, touched with the hands, or detected by instruments.*

5. *The perfection of any logical idea is the unattainable limit toward which contradictory matter constantly tends but which it will never reach.*

Thus, except for its own subjective whims and illusions, what could atheism use to counter the scientific law of the objective existence of ideas and the Ideal World? It can only use its absolutely unproven initial assumption that "there is nothing in the world except for matter in motion," despite all the scientific evidence.

But then we might logically ask: where is the scientific evidence for atheism? The answer is simple, namely, there is no scientific evidence for "scientific" atheism whatsoever! It only has its convenient *initial assumptions,* which hundreds of millions of simple people *have been forced to believe blindly*, even though it is logically impossible to believe in them. In reality, the natural sciences have definitively proven the objective existence of weightlessness and such invisible categories as radio waves, the human mind, the laws of nature, and so forth.

Nevertheless, the atheist usually rejects any ideal categories in the following terms:

"Objective ideas? Ideal spirits? Absolute God? What in the world are they? Where are they? If they really do exist, then show them to me! Let me see them with my own eyes! Let me touch them with my own hands! You cannot do that? Aha! That means that there are no such things as an ideal spirit or an Absolute God!"

I have a short and simple answer for all these questions: "Do you have a mind?"

"Oh yes! Of course I do," the atheist exclaims.

Then I launch my counterattack.

"A mind? What in the world is that? Where is it? A mind is not a brain, it is rather reflected by the brain. If you really do have not only a brain but also a mind, then show it to me! Let me see it with my own eyes! Let me touch it with my own hands! You cannot show it to me? Aha! That means one of two things: either your atheistic logic is completely wrong, or you don't have a mind at all!"

"No, no! Of course I have a mind and my logic is completely correct!" the atheist responds. "It's true that we can't see the mind with our eyes, touch it with our hands, or detect it with instruments, but we can surmise (infer!) its existence from my visible and perceptible behavior."

Consequently, the atheist asks me to recognize the existence of his invisible and weightless mind not by means of direct empirical perception but as a result of inferences based on this perception. By perfect analogy, we can perceive the undeniable fact of the existence of an Absolute God and objective ideas not as a result of direct physical contact with them but as a result of inferences based on reliable scientific facts.

The human mind is just one of many classic examples which clearly persuade us that not all reality is weighty and perceptible and that not every reality can be directly seen with the eyes, heard with the ears, touched with the hands, or detected with instruments. Other examples of weightless and invisible objective realities include the laws of nature, the ideal program for the universal development of matter (the complete collection of all natural laws), the genetic program (the semantic contents of genetic codes, and not the codes themselves), real time (not the movement of a clock!), and so forth.

The notorious atheistic principle that "the world contains nothing which I cannot see with my eyes, hear with my ears, touches with my hands, or detect with instruments" has been scientifically obsolete for a long time now. On the contrary, the modern natural sciences adhere to the following principle: everything which has been definitely proven by science is worthy of recognition.

# =7=

# THE THEORY OF THE EXPANSION OF THE UNIVERSE

> *If the Universe did not expand and were infinite, the temperature in it would be so high that the formation of even the simplest molecular compounds would be highly improbable.*
>
> Joseph Shklovsky

So-called "scientific" atheism and "dialectical" materialism have constructed their antireligious theories on the implausible initial premise of the "eternity and infinity" of the Universe. No one has ever provided theoretical proof or experimental confirmation of this initial assumption. Hence it would be quite appropriate to ask: where are the scientific proofs of atheism? The answer is simple: "scientific" atheism has no scientific proofs nor could it! "Scientific" atheism has only its convenient initial premises which *hundreds of millions of people have been compelled to believe blindly*, even though it is logically impossible to believe them, because the natural sciences have definitely established the opposite. Atheism has made a desperate attempt to counter scientific proof with "criticism" of any scientific theory which directly or indirectly persuades us of the boundedness of the Universe in time and space. Our attention is drawn to the fact that this "criticism" contains no science whatsoever and merely consists of unfounded, trite, and high-flown pronouncements.

Let us give a few examples.

### First example (From a speech by A.A.Zhdanov):
Modern bourgeois science is supplying the priests and the theists with new arguments which must be mercilessly exposed. Without

comprehending the dialectical progression of knowledge or the relationship between absolute and relative truth, many of Einstein's followers have transferred the conclusions of their investigation of the laws of motion of a finite, limited region of the Universe to the entire infinite Universe and have attempted to argue the finite nature of the world and its boundedness in time and space, and Milne, the astronomer, has even "calculated" that the world was created 2 billion years ago. These British scientists may be quite aptly characterized in the words of their great countryman, the philosopher Bacon, who said that, they have transformed the futility of their science into a slander against nature.[1]

It would have been quite appropriate to ask Zhdanov whether the scientist or he himself was the one slandering nature. Who needed to "mercilessly expose" the scientifically proven religious truth of the expansion of the Universe and for what purpose? Who made it necessary and for what to send scientists who not only believed but were convinced of the existence of God into Siberian exile and to cruel deaths? But the radio only said what it said and did not answer any questions. In the heat of his atheistic ardor, A.A. Zhdanov, the fiery orator and statesman, didn't realize and didn't know that only 20 years into the future atheism would have to surrender to this purely religious scientific theory of the expanding Universe under the pressure of undeniable scientific facts.

### Second example:
"The proliferation of a variety of idealistic theories of the expanding Universe in foreign literature has evoked sharp criticism of these theories on the part of materialist scientists. The idea of the expansion of the Universe has quite rightly been assessed as anti-scientific and promoting theism."[2]

Remember that materialist scientists considered the theory of the expansion of the Universe anti-scientific. This definitely implies that if the theory of the expansion of the Universe is in fact scientific, then materialism is in fact unscientific.

### Third example
One attempt to refute the notion of the infinity of the world is the idealistic theory of the expanding Universe . . . Idealist philosophers and astronomers have concluded that at one time the entire Universe was concentrated in an extremely small finite volume, a kind of primordial atom, but that at some point in time it began to expand suddenly, which was accompanied by the expansion of space, which was originally infinitesimally small. This theory was accompanied by the declaration that God had created this primordial atom and that his will had caused it to expand. This reactionary, blatantly theistic theory of the expanding Universe cannot withstand the slightest scrutiny.[3]

So what now? Now we must say that the truth is still the truth, regardless of how atheists or materialists might criticize it. Subsequently the materialists softened their criticism of the scientific model of the expanding Universe, then shamefully covered their eyes and remained silent for a while. Now unanimously, as if on command from a puppeteer, they solemnly declare this theory to be scientific.

For a long time the scientific model of the expanding Universe was subjected to slanderous attacks by materialism and atheism. They declared this theory to be "blatantly religious" and called all of the scientists who supported the theory "phony scientists." They did not spare anyone and even hurled insults at the theory's author, Albert Einstein, despite his world renown. But the atheists did not succeed in their attempts to destroy the theory of the expanding Universe with juicy anti-religious pronouncements. Their tried and true method of attacking scientific logic with emotionalism failed. The fiery orator who shouted angrily at the entire world using his powerful radio transmitters, university podia, newspapers, and journals proved to be wrong. The humble scientist who calmly spoke the truth proved to be right. And the orator could only shout down the truth temporarily. The truth always prevails in the end.

Alexander Isaakovich Kitaygorodsky, the prominent physicist and tireless opponent of the advocates of scientific religion, was among the first materialists who had to surrender unconditionally to the blatantly religious theory of the expanding Universe. He wrote:

> A study of the Doppler effect in the spectra of stars in different galaxies has definitely proven that all of them are running away "from us." Furthermore it has been demonstrated that the speed of flight of a galaxy is directly proportional to its distance "from us." The most distant galaxies which physicists can see are traveling at speeds approaching half the speed of light... Einstein's model of the Universe proposed in 1917 is the natural corollary of his so-called general theory of relativity. Einstein, however, did not assume that a closed Universe had to expand. This was demonstrated in 1922-1924 by the Soviet scientist Aleksandr Aleksandrovich Fridman (1888-1925). It turned out that the theory required either the expansion of the Universe or alternating expansions and compressions.[4]

As you can see, materialism was ultimately compelled to recognize the theory of the expanding Universe. But in addition to recognizing the theory as scientific, it has also defended its own preeminence. It turns out that "the idealistic theory of the expanding Universe" was developed not by the idealist Albert Einstein, but by the Soviet scientist (i.e., a materialist!) Aleksandr Aleksandrovich Fridman. So that's how it is! But what are the implications of the fact that a materialist developed an idealistic theory? Does it follow that idealistic theory has became a materialistic theory? Not at all! The theory of the expanding Universe was, is, and will continue to be an idealistic theory. The fact that a materialist was the first to develop a particular idealistic theory merely demonstrates once again that if a materialist thinks objectively and wants to remain a scientist, sooner or later he will inevitably proceed to idealism. But a scientific materialist cannot call himself an idealist as long as he gets his paycheck from an atheist who could not care less about the objective truth.

The futility of atheism's open warfare against the scientific model of the expanding Universe gradually became clear even to the atheist scientists. Hence the only recourse left to them was to recognize this theory solemnly and loudly, as if nothing had happened. And right now the materialists not only recognize the theory of expanding Universe but claim it as their own. Hence it would be quite logical for atheism to admit its defeat and depart the world arena. But instead, atheism has begun to adapt this scientific theory of the expanding Universe, which it previously called "openly religious," to its own fundamental dogmas. This example demonstrates how atheism tries to use the conclusions of modern scientific theory after first perverting and distorting them. But how is it possible to adapt a "purely religious" theory to the basic dogmas of atheism? After all, the scientific theory of the expanding Universe definitely admits that billions of years ago the newborn Universe was no larger than an elementary particle. Hence it does not leave any room for atheist fairy tales concerning the "infinity and eternity" of the Universe. And atheism could not exist as an anti-religious doctrine without these concepts of "eternity" and "infinity." Hence atheism has now been compelled to formulate its own atheistic version of a scientific (purely religious) theory.

Back in 1976 (when the Soviet Communist Party and totalitarian atheism were at the zenith of their power), Corresponding Member of the Soviet Academy of Sciences Joseph Samuilovich Shklovsky formulated this version as follows:

> Approximately 12 billion years ago, the entire Universe was concentrated in a very small area. Many scientists believe that at

that time the density of the Universe was about $10^{14}$ to $10^{15}$ g/cm$^3$, i.e., roughly the same as that of an atomic nucleus. And still earlier, when the Universe was infinitesimal fractions of a second old, its density was much greater than nuclear density. Simple speaking, at that point in time the Universe was a single gigantic "drop" of supernuclear density. For some reason this drop became unstable and exploded. We are observing the aftermath of this explosion right now as the dispersion of the system of galaxies.[5]

From this quote we can see that if a scientist is paid by the leaders of atheism, he is physically compelled to come to the defense of atheism. But at the same time, the noble calling of a prominent scientist forces him to be extremely cautious in his statements. Hence Joseph Samuilovich has to represent the newborn Universe as some sort of mysterious "drop with supernuclear density," even though this description is quite shaky from both a scientific and an atheist point of view. In fact, if the Universe was a drop at one time, then it is not infinite and eternal. If matter is not infinite, then it is also not eternal. And if matter is not eternal, this means that it had a beginning. If matter had a beginning, this means that matter was created by some kind of nonmaterial (otherworldly) force. Thus, all it takes is one step to get from Shklovsky's atheistic notions to scientific religion.

That is why many run-of-the-mill scientific atheists have categorically stated that the primordial Universe was some kind of particle with an infinitely great density.

Atheism, which has loudly and persistently proclaimed its formal "scientificness" to the entire world, could in reality not exist without such implausible concepts as the "infinity and eternity" of matter. But the scientific model of the expansion of the Universe leaves absolutely no room for atheist fairy tales about the "infinity and eternity" of the Material World. Hence that is why atheism has been compelled to hastily concoct a new fairy tale about "the infinite density of the zero volume" of the primordial Universe. It is sad to see a white religious theory sewn up with the black threads of atheism.

Instead of science, totalitarian atheism has asked hundreds of millions of people to believe blindly in its new fairy tale about the "infinite density of the zero volume" of the newborn Universe, even though it is logically impossible to believe it for the following reason:

We know that if the mass of a physical body is greater than twice the mass of our sun and if its density is greater than that of an atomic nucleus, then there are no natural forces capable of halting its catastrophic gravitational compression, which will continue until the body vanishes in a black cosmic hole. It has been theoretically proven that it would be completely impossible for this body to expand.[6]

*If matter were uncreatable and indestructible, as the atheists claim, then the mass of all of our current stars and galaxies, which would exceed the mass of our sun not by a factor of two, but by a factor greater than $10^{21}$, would have been concentrated in the zero or extremely small volume of the primordial Universe. Hence if matter were uncreatable and if the newborn Universe were a substance condensed in a small volume to supernuclear density, then the expansion of the primordial Universe would have been impossible. Moreover, this kind of superdense primordial Universe would inevitably continue to condense until it completely vanished in its own "black hole" without leaving anything material behind it, not even pure energy.[7] There are no physical forces in nature which would be capable of preventing this kind of catastrophic disappearance of the Universe.*

The expansion of a zero point with an infinitely high density would be even more impossible.

But the fact of the expansion of the Universe is obvious. From this we will conclude that matter is creatable, that the Universe was born and began to expand from an ideal zero or zero point with no physical dimensions, no physical volume, no physical energy, no mass, no weight, no density, and no physical attributes at all.

This signifies the collapse of atheism's new fairy tales of the "infinite density of the zero volume" of the newborn Universe. Now atheism cannot refer to the "inadmissability" of scientific laws because the primordial Universe had no supernuclear density whatsoever. But how could atheism counter these irrefutable scientific facts now? It has nothing except its own whims. The only decent solution for atheism now would be to sign its unconditional surrender to religion with dignity and honor.

But instead, atheism is still repeating its old fairy tales about the uncreatability of matter, and hundreds of millions of simple and naive people are still compelled to believe these fairy tales blindly, even though it is logically impossible to do so, because the expansion of the Universe would be impossible without an increase in its so-called vacuum space, which in practical terms is a weightless and invisible physical ocean of negative energy.

The expansion of the physical space of the Universe without a corresponding increase in its negative energy would be impossible in exactly the same way as it would be impossible for the Atlantic Ocean to increase in volume without a corresponding increase in the volume of water in it.

Consequently, the expansion of the Universe must be accompanied by a continuous increase in negative energy. According to the law of conservation, a continuous increase in the Universe's negative energy must be accompanied by an influx of the same quantity of positive energy so that the algebraic sum of positive and negative energy always remains constant and equal to zero.

By perfect analogy, the catastrophic contraction of the Universe would be impossible without a decrease in its vacuum space. This means that the gravitational collapse of the Universe would have to be accompanied initially by a continuous or periodic decrease in the negative energy of vacuum space and ultimately by its total and utter disappearance. According to the law of conservation, the annihilation of the negative energy of vacuum space would be impossible without the annihilation of the same amount of positive energy in the collapsing Universe.

*Thus, a necessary condition for the expansion of the Universe is the creation of energy from nothing, and an inevitable effect of the gravitational collapse and catastrophic contraction of the Universe would be the complete disappearance of all its energy.*

If energy were uncreatable, then the physical space of the Universe, which is a raging ocean of negative energy, could not expand. But the fact that the Universe and its space are expanding is obvious. Consequently, energy is creatable.

To assert that matter is uncreatable and at the same time recognize the expansion of the Universe is, in the final analysis, equivalent to "squeezing" the entire Universe, including our earth, the sun, all of the stars, and all of the galaxies into a single primordial point with zero volume. These sorts of absurdities are not even found in fairy tales. The only place you may find them is in atheist "science." As a matter of fact, any fairy tale limits itself to squeezing a big genie into a small bottle. But no teller of fairy tales could permit himself a fantasy such as squeezing an "infinite mass into a zero point." Atheism not only permits itself the luxury of telling such absurdities to people, but it even calls them "scientific." And it does all this at a time when religion modestly calls its scientific truths "faith." The infinite density of a zero point is an abstract, i.e. imaginable, but practically impossible category which atheism attempts to use to salvage its fairy tale of the "uncreatability and indestructibility" of matter.

Thus, the scientific theory of the expanding Universe is recognized without reservations even by atheists.[8] We may concisely formulate this theory as follows:

1. *The Universe was born and began to expand approximately 12 to 14 billion terrestrial years ago from an ideal zero or zero point which contained no galaxies, no stars, no sun, no earth, no physical energy, and no matter whatsoever. We will call this primordial ideal point the first white cosmic hole. The Universe is still expanding now. It is spherical in shape, and its spatial extension at present does not exceed $3 \cdot 10^{23}$ km.*

2. *It has been scientifically proven and experimentally confirmed that the Universe is not eternal in time or infinite in extension. Science begins at the point where and when the atheistic dogmas of the fictitious "eternity and infinity" of the Universe leave off.*

# =8=

# THE ABSOLUTE GOD AND THE RELATIVE WORLD

## THE BIBLICAL MODEL

"*I*n the beginning God created the heaven and the earth.*"[1] In this case the word "heaven" means space and time, and the word "earth" means everything that exists in space and time.[2] The significance of the first verse of the Bible lies primarily in its emphasis on the absolute primacy of God and the relative subordinacy of the entire world, both the physical and the nonphysical. According to the Bible, "In the beginning God created" space, time, and everything that exists in space and time. This means that God existed outside of any space and outside of any time even when there was no time, no space, no physical bodies, no earth, no sun, no galaxies, no Universe, and no world whatsoever. This leads to the unambiguous conclusion that the Bible portrays God as an absolute category instead of a material category, because God came before the creation of, not just matter, but the entire Relative World.

According to the Bible, God is an eternal, bodiless, and absolutely perfect category which has no physical attributes and no biological organs. God is not a biological being. God is the creator of living beings. Even space and time are a product of the creative endeavors of the Absolute God, not the realm of God's own existence.

Atheism has shamelessly slandered the Holy Bible by depicting God as a "bearded old man."[3] But then we might properly ask: what does modern science have to say on the subject?

## The Modern Scientific Model

By its very nature, any intelligent idea is more perfect and useful than unintelligent matter, even though an idea cannot be seen, heard, touched, or detected by instruments. Nevertheless, this sort of objective idea is still a relative category (but not an absolute category!) because we can always apply such terms as more perfect or less perfect, more refined or less refined, better or worse, more or less, earlier or later, closer or farther, and so forth to it.

Einstein's theory of relativity shows that there is nothing absolute in the physical world nor can there be. This means that, to us, weightless space and invisible time as well as physical bodies and weighty and visible substances are relative categories, because we can always compare them in such terms as more or less, closer or farther, earlier or later, and so forth.

According to the fundamental law of nature, however, *relative matter* cannot exist in space and time without its *absolute opposite*, which exists outside of any space or any time, and for which there are no comparative categories; no better or worse, no more or less, no closer or farther, no earlier or later, and so forth.

A person is a person inasmuch as that person may be distinguished in some way from all other persons. If the outward appearances of some set of persons were exactly alike, then we could no longer distinguish them from each other and all of them would visually seem to be one and the same person to us. If these same people had the same ideas, same mentalities, and same spiritual characters and so forth at the same time, we could no longer distinguish them in our minds as well as with our eyes.

If in addition to all these other attributes, the will of one person were exactly the same as the will of another person, these two persons would become one not only in our subjective consciousness but in objective reality. By perfect analogy, if any set of absolute categories were absolutely identical, these categories would combine to form a single absolute category. If there were many absolute categories and all of them differed from one another in just the slightest degree, then they could be compared and would thus cease to be absolute. Hence an absolute category can be one and only one unique category. To put it in simpler terms, it would be just as impossible to have two absolute categories as it would be to have two world chess champions at the same time. We also could not call an absolute category absolute unless it was absolutely perfect. Hence an absolute category can only be an absolutely perfect and unique category.

By its very absolute essence, an absolute category may not be dependent on any relative category. If this were not the case, the absolute category would no longer be absolute. At the same time, according to the theory of relativity, space and time are relative categories.

If absolute perfection were to exist in space and time, then it would be dependent on the relative space-time continuum and would thus cease to be absolute perfection, because any space and any time are always relative categories. Hence we should look for an absolute category outside of any space and any time.

The space-time continuum is a realm of the motion, change, and development of reality. If the very existence of reality did not require any motion, change, or development, there would be no need for any space or time whatsoever.

The more refined reality is, the less its volume is and the closer it is to absolute perfection. The existence of a more refined category requires less motion, change, and development. Hence more perfect categories have less of a need for the space-time continuum than less perfect categories. And at the limit, an absolutely perfect category requires no space or time for its own existence.

By its very nature and by virtue of its absoluteness, absolute perfection constitutes the height of any perfection to the extent that it requires no motion or development. Any space and any time are realms of relative motion and development. Hence absolute perfection does not need any relative space-time and exists in absolute eternity outside of any space and time.

Thus, a unique and absolutely perfect category is a necessity without which relative matter and the Material World could not exist in general. From daily life we know that the Material World exists in reality and objectively, outside and independently of any subjective (human) desires. Consequently, an absolutely perfect category also exists in objective reality, regardless of what we may think. If we admit the objective existence of relative matter in space and time, then according to the fundamental law of nature, we must equally admit the objective existence of its absolute opposite outside of any space and outside of any time. We use the word God for this unique and absolutely perfect category which exists in absolute eternity outside of any space and time.

Either we admit the scientifically proven fact of the existence of an Absolute God or we must reject the fundamental law of nature, whose truth is completely obvious, in spite of the scientific facts. We have no other alternatives.

In order to exist, the Absolute God needs no extension of space and time, because any extension of space and time is always a relative category.

The absolute person, the absolute will, the absolute intelligence, the absolute idea, the absolute truth, absolute eternity, and so forth are all real and this-worldly manifestations of the otherworldly essence of the one and indivisible Absolute God.

The relative mind can comprehend the relative truth, but not the absolute truth. Our minds are relative, not absolute, categories. That is why we will never be able to grasp or comprehend the full truth of the Absolute God. We only know the relative truth that God exists in absolute eternity outside of any space and outside of any time. But we cannot even comprehend the significance of this absolute eternity, where there is no earlier or later, no closer or farther, no better or worse, where zero merges with infinity, and where the absolute will, absolute intelligence, and absolute idea are all embodied in the absolute being of the one and indivisible God.

God is an absolute (and not relative!) category whose perfection is superior to any other perfection. The absolute perfection of God could not be achieved by anyone at any time. Any relative idea (or "ideal spirit") is an objective reality of a much lower quality than the Absolute God and of a much higher quality than physical energy. If we call a relative idea (or "ideal spirit") "perfect" by comparison with mindless matter, then we would have to call the same relative idea "primitive" by comparison with an absolute category.

An ideal spirit created by God has a beginning, even though it may be eternal and immortal in the future. An ideal spirit with a beginning but no end could not become the Absolute God with no beginning and no end even if it were to develop perpetually, no matter how close it came to an absolute perfection. In order to become the Absolute God, an ideal spirit would have to lose its "beginning," which would be impossible.

An ideal spirit eternally striving for an absolute perfection in relative space and in relative time could never become the Absolutely Perfect God, which exists outside of any space and outside of any time. There is a space-time boundary between them which is impassable for any relative category. In order to become the Absolute God, an ideal spirit would have to "depart the realm of space and time," but without space and time an ideal spirit could not only not develop but would not exist at all. The absolute perfection of God is that unattainable limit toward which any other intellectual (intelligent) category strives continuously and eternally in the process of it development, but never reaches it.

If the Absolute World of God were to develop in time, then it would depend on relative time and would thus cease to be absolute. It is not the Absolute God which develops in relative time, but relative time which is the product of the creative activity of the Absolute God. Hence the World of the Absolute God is the absolute beginning of any time.

According to the fundamental law of nature, the relative space of the physical world could not exist without its absolute opposite, which has no closer or farther, no earlier or later, no less or more, and so forth. We could arbitrarily use the term "absolute space" to refer to this absolute opposite of relative space. But in our human minds, this kind of "absolute space" with no closer or farther would no longer be a space. It would be a point, all of whose spatial dimensions are equal to absolute zero.

If an absolute category were to have any (even the smallest) dimensions, then it would have to exist in relative space and would thus cease to be an absolute category. Hence to us (as representatives of the Relative World) the World of the Absolute God is a point where all dimensions are equal to absolute zero.

If this kind of absolute point were to move in relative space or relative time, it would be dependent on relative space and would thus cease to be an absolute category. It is not the absolute category which moves in relative space but the relative space which expands relative to the absolute point. This means that the World of God is the absolute origin (center-point) of any space.

It is not God which depends on space and time, but space and time which depend on God. God's Intelligence is so great that it grants God absolute free will. This means that God, who exists outside of any space and any time, has an infinite number of degrees of freedom even in an absolutely zero space. Hence God can create an infinitely large space-time realm with an infinitely large number of dimensions around any point, all of whose dimensions are equal to absolute zero.

Absolute God, who exists in absolute eternity outside of any space and any time is the absolute center of all existence (both material and ideal). In the geometric meaning of the word (in very simplified terms!) this center may be represented as a point, all of whose space-time dimensions are equal to zero.[4]

From the absolute center of the world, at which God exists eternally outside of any space and any time, an ideal and infinite space-time continuum with an infinite number of dimensions is expanding in all directions at an infinitely high velocity. These dimensions of relative space and time are not only a secondary product but a relative field of creative activity of the Absolute God, whose will is completely unlimited and free in the absolute sense of the word.[5]

It has been scientifically proven that the three-dimensional physical space of our Universe originated and is expanding in a multidimensional ideal space.[6] But then we might naturally ask whether this Universe began all by itself or was created by God.

Throughout its entire history, intelligent humanity has never come up with a greater absurdity than anti-scientific atheism, which has intentionally and quasi-scientifically tried to distort the truth and thus cast a "shadow on a clear day." But the truth of the creation of the Material World by an Ideal God can be proven very easily by means of the three scientific laws of nature, which have been many times confirmed by theory and practice.

According to the first law, no material system can exist eternally: it is born, develops, reaches the highest stage of its development, inevitably accumulates error, ages, and dies. The Universe in which we live is a material system and therefore cannot exist forever. If the Material World consists of a number of physical Universes, this does not mean that it ceases to be a material system, and therefore it must have a beginning and an end. So what grounds did atheism have for exempting the Material World from this general law? Absolutely none, except for its own stubborn whims!

According to the second law of nature, the cause of the formation or birth of a given material system always lies *outside* the system. No material system can originate and develop on its own without an external cause, outside help, or without an initial stimulus. For example, children cannot be born without parents, and a watch will not run unless it is manufactured, wound, and so forth. That is why the Material World could not originate and develop by itself without some otherworldly (nonmaterial) intervention. Consequently, the Material World was created by a nonmaterial being. But by whom? The answer to this question is provided by the third law of nature, which states that matter in the Universe develops in a highly purposeful way according to a definite and predetermined program. But a purposeful program can only be the product of intelligent creativity. Consequently, the Creator of the Material World possesses an extraordinarily high intelligence. And we use the word God to refer to this highly intelligent Creator of the Material World.

To believe that the Universe originated and is developing in a purposeful way on its own without an intelligent Creator (without God) is equivalent to saying that a television set sprung up on its own, without any intelligent inventor, without a designer, without any creator, or without any engineering idea. There are many other scientific proofs of the absolute primacy of God and the relative subordinacy of the Material World which you can find, for example, in the book "*Миры*".[7]

According to the theory of relativity, matter contains nothing absolute. Hence an absolute category cannot be a material category. God, as the absolute opposite of relative matter, does not contain anything material. Thus, the first chapter of the Bible has been completely confirmed by the results of modern science: God is a unique, absolutely primary, and absolutely perfect ideal category with no material attributes whatsoever.

But then we might ask how the Absolute God, all of whose spatial dimensions are equal to zero, has created and ruled such a vast world. In order to get a better grasp of this question, it would be useful to consider the following example.

Let us assume that a police cruiser is patrolling a sector with a total perimeter of 10 kilometers. Let us imagine that the cruiser's speed can be increased indefinitely, to infinity, even though this is impossible in reality. *Infinite speed* means that speed at which a real object will cover any distance imaginable in the shortest possible time imaginable.

If the cruiser is traveling at a speed of 20 kilometers per hour, it would pass by us every 1800 seconds. You can see with your own eyes that the faster the cruiser travels, the more often it will pass by you. But if the cruiser's speed were increased to 500 kilometers per second, you would no longer be able to determine the visual frequency of its appearance. No matter where you looked (on its trajectory), it would always seem as if the vehicle were "frozen" right where it is. That is why electric lights and the frames of a movie look uninterrupted to us.

The dimensions of the cruiser would begin to shrink noticeably if its speed approached the speed of light. At the speed of light the length of the cruiser with the policemen sitting in it would appear to be zero in your eyes. If the cruiser's speed were to increase even more, you would not be able to see anything, because the direct transmission of a material signal from the world of faster- than-light speeds to our world of slower-than-light speeds is physically impossible. Your vision would completely exhaust its capabilities. But you also have logic in your corner, which would allow you to calculate how many times the vehicle passed by you as long as you knew its speed. But if the cruiser's speed were to approach infinity, even your rather high intelligence would completely exhaust itself and you would be unable to answer the question of where the cruiser was at any particular instant. The cruiser has literally disappeared before your very eyes into the otherworldly Ideal World of infinitely high velocities. Hence the cruiser is everywhere (on its trajectory) at any given instant, not only in the eye but in your mind. But at the same time it is nowhere.

If a point is moving in a circle at a finite velocity, you will always be able to determine the location of this point at any given moment in time. But if the point's velocity were to become infinitely great, then you would be unable to locate it even mentally. At the same time it is everywhere, because in any infinitesimally small interval of time it will "travel" around the circle an infinite number of times. Because of the point's infinite speed, the concept of "nowhere" becomes equivalent to the concept of "everywhere."

By perfect analogy, the ideas of God, which possess an infinitely great velocity, "travel" around the entire world an infinite number of times in any infinitesimally small interval of time, even though we are incapable of detecting these ideas physically. Under certain conditions the human brain is capable of duplicating an objective idea. But even in this case we cannot precisely determine the exact location of this subjective duplicate of an objective idea in our brains.

These mental experiments should persuade us that the atheist's usual question of "Where is God?" is irrelevant, because God is an absolute category who exists in absolute eternity outside of any space and any time, and the ideas of the Absolute God, which govern the entire world, are propagated in all dimensions of relative space simultaneously at an infinitely high velocity. But if we were still asked this question in this primitive form, then our answer to it would be simple: God is everywhere at any moment in time, but cannot be seen anywhere at any time. God cannot be detected physically anywhere, even though God is present everywhere as an invisible and intangible ideal category. Even though in the absolute meaning of the word God exists in absolute eternity, outside of any time, outside of any space, and is in a state of absolute rest, God's influence extends to any point of the Real World instantaneously at any moment in time at an infinitely high speed.

While the velocities of weighty and visible bodies usually do not exceed the speed of sound, the velocity of pure energy usually does not exceed the speed of light. God, however, is able to spread ideal information throughout the entire world at an infinitely high speed so that it is impossible to say where God is.

God uses transitory-absolute information to control and rule the entire Relative World. Any relative information is definitely and unambiguously subordinate to this information. It is the infinitely great speed of the propagation of ideal information only for the unique and absolutely perfect God, and only in our subjective (not overly "refined") consciousness does the concept of "nowhere" become equivalent to the concept of "everywhere," absolute zero merges with the endless spaces of the world, and absolute rest merges with absolute motion. What to us is absolute eternity is an infinitesimal fraction of time to God. God's absolute perfection lies first and foremost in the fact that God can engage in incredible creative activity, rule the infinite world, and be in a state of absolute rest at the same time.

Thus, God is an absolutely primary reality which combines absolute zero with infinity. The combination of all other forms of objective reality is the Relative World, which was created by the Absolute God, is subordinate to God, and is secondary in relation to God.

On the basis of the latest advances in the natural sciences, we may summarize the law of the absolute primacy of God and the relative subordinacy of the world as follows:

*1. In the beginning God created space, time, and everything which moves and develops in space and time ("In the beginning God created the heaven and the earth.") This means that God is absolutely primary, and that the entire world created by the Absolute God, both the material and ideal worlds, is relatively secondary.*

*2. God is the absolute opposite of relative matter and, therefore, does not contain any physical organs or material attributes, because all matter and* any physical organs are only the external (relative) products of the creative endeavors of the Absolute God instead of an internal part of this God. God has no physical body, no biological brain, and no material eyes, ears, hands, feet, beard, and so forth. God cannot be seen with the eyes, touched with the hands, or detected by instruments, just as the human mind cannot be seen with the eyes, touched with the hands, or detected by instruments.[8]

3. God is a unique, absolutely perfect reality which exists in absolute eternity outside of any space and time. Space and time are merely the relative products and fields of the creative activity of the Absolute God, and not the realm of God's own existence. At the same time God is omnipresent and everlasting reality, because God controls and rules the entire Relative World by means of signals and information communications, which may at any time be propagated to any point in the real world instantaneously at an infinitely high speed. God never had a beginning, and God will never have an end in the absolute meaning of the word. God existed when there was no space and time. Hence any atheistic questions, such as "Who created God?" or "Where is God?", are meaningless and do not have any scientific contents.

4. The Absolute God who exists in absolute eternity outside of any space and outside of any time is the absolute center of all existence (material or ideal). In the geometric sense of the word (in very simplified terms!), this center may be arbitrarily and only arbitrarily represented as a point, all of which space-time dimensions are equal to absolute zero.[9]

God is the absolute origin (center) of space, time, and every objective reality which moves and develops in space and time.

5. An ideal and infinite space-time continuum with an infinite number of dimensions is expanding simultaneously in all directions at an infinitely high velocity from this absolute center of the world. These dimensions of relative space and time are both a secondary product and relative field of creative activity of the Absolute God, whose will is limited by nothing and is free in the absolute sense of the world. It has been scientifically proven that the three-dimensional physical space of our Universe originated and is

*expanding in this multidimensional ideal space.*[10]

6. *The scientific concept of the Absolute God, who exists eternally outside of any space and outside of any time, is perfectly consistent with the Biblical concept of God who in the beginning created "the heaven and the earth," i.e. space, time, and everything which exists in space and time.*

Hence, except for its own whims and illusions, what does atheism have to say to counter the scientific law of the absolute nature of God and the relativity of the world? All it has to counter these scientific proofs are its completely unfounded initial assumptions that objective ideas, "ideal spirits," and an Absolute God do not exist simply because we cannot see them with our eyes, hear them with our ears, touch them with our hands, or detect them with instruments.

But then we might logically ask where are the scientific proofs of atheism? The answer is simple: "scientific" atheism has no scientific proofs whatsoever! It only has its convenient subjective assumptions which hundreds of millions of people have been compelled to believe blindly, even though it is logically impossible to believe them. In fact, the natural sciences have conclusively established the objective existence of weightless and invisible categories such as radio waves, the human mind, the laws of nature, and so forth.

The atheists have reserved their fiercest attacks for the first paragraph of the Bible, which asserts the absolute primacy of God and the relative subordinacy of the entire world. In its "fundamental law of philosophy," totalitarian atheism has forthrightly declared that "Matter is primary, and the idea is secondary." At the same time Lenin, the leader of totalitarian atheism, categorically objected to the Absolute God and the ideal human spirit on behalf of atheist science in the following terms: "There is nothing in the world except for matter in motion."[11]

But if the world truly contains no ideas and nothing at all except for matter, how is it possible to believe in the secondary nature of ideas, which do not exist? If ideas are secondary, this means that they exist. And if ideas exist, how can one believe that "there is nothing in the world except for matter in motion?"

Nevertheless, this phrase has focused the attention of hundreds of millions of people on the concept of "matter" in such a way that their imaginations cannot escape it even now, after the utter scientific collapse of totalitarian atheism.

It would be impossible to fly to the Moon without breaking away from the earth. By perfect analogy, you could not penetrate the essence of the concept of "idea" as long as your thinking cannot escape the confines of the concept of "matter." And if you want to mentally enter "another" ("ideal") world, then you must try to mentally escape the confines of "this" (material) world. Otherwise you will never be able to "see" the Absolute God, even in your own mind.

If a deaf person cannot hear, this does not mean that there is no sound or music in the world. If a blind man cannot see, this does not mean that daylight does not exist. And if a limited mind cannot comprehend the majesty of God, this does not mean at all that the Absolute God does not exist.

# =9=

# THE THEORY OF THE EVOLUTIONARY UNIVERSE

*At present, we may consider the fundamental proposition
that the Universe is evolving, and is evolving quite intensely,
a completely proven fact.*

Joseph Shklovsky

Totalitarian atheism has intentionally indoctrinated millions of people with the obviously false idea that the scientific theory of evolution, and even the word "evolution" itself, discredit the Bible. But then we might logically ask if this is really the case. And in fact, the ancient Hebrew language had no such word as "evolution." But this in no way implies that the word "evolution" in and of itself contradicts the Bible. Many of the people living today, not to mention ancient people, do not know the exact meaning of this scientific term. According to the scientific definition, any non-revolutionary gradual development in which each subsequent stage is qualitatively different from the preceding stage is called *evolution*. And this is exactly what the Bible talks about when it says the world was created in six stages (on the six Biblical "days"): each subsequent stage of the creation of the world was *qualitatively* different from each preceding stage.

Even the atheistic and scientific definitions of the term "evolution" are not inconsistent with the Bible in any way. By totalitarian atheism's definition, "any change involving a gradual quantitative change of that which exists is called *evolutionary*."[1] "As they accumulate gradually, at a certain point quantitative changes alter the scale of the object and cause radical, qualitative changes."[2] And this is exactly what the Bible says: the transformation of the nonexistence of the Universe into its existence was accompanied by steady (gradual) quantitative changes which eventually became qualitative changes in six different stages.

But then we might rightly ask the following question: if the term "evolution" does not refute the Bible *a priori* or *per se*, then does not the scientific theory of the evolutionary Universe contradict the Biblical model of God's creation of the world?

The natural sciences are characterized by two conflicting schools, namely the theory of the steady-state Universe and the theory of the evolutionary Universe.

In 1948, Fred Hoyle, the prominent British astrophysicist and two other scientists advanced their "theory of the steady-state Universe" which states that the basic qualities of the Universe have presumably never changed and that after tens and hundreds of billions of years it looks the same as it did hundreds of billions of years ago. This implies that pure energy, plasma, liquids, solids, and living beings have presumably existed, exist, and will continue to exist for eternity in the Universe. Individual stars and galaxies are born and die, but the world of stars and galaxies as a whole has remained unaltered for eternity. All of this inevitably leads to the conclusion that the Universe is presumably eternal and infinite, and not evolutionary. If the Universe were not evolutionary but instead was eternal and infinite, it would have no beginning or end and consequently would not be the product of any creative endeavor. And this is exactly what atheism needs to exist. Hence it is the *steady-state theory, not the evolutionary theory, which is purely atheistic.*

But recent advances in the natural sciences have completely discredited the steady-state theory. First of all, scientists demonstrated that the Universe began to expand from a zero point to its current dimensions. Subsequently they established that a necessary condition for the expansion of the Universe is the creation of energy from nothing, meaning that as the Universe has expanded, matter has arisen from nothing step by step. The spatial expansion of the boundaries of the Universe would inevitably have to be accompanied by a continuous increase in the amount of matter in the Universe. According to the dialectical law of the transformation of quantitative changes into qualitative changes, this continuous increase in the amount of matter in the Universe should periodically be accompanied by qualitative changes in the Universe. Therefore, the Universe has not only increased in quantitative terms, but it has also changed qualitatively. Any gradual change in quantity which is accompanied by incremental improvements in quality is commonly known as *evolutionary development*.

According to the evolutionary theory and in full accord with the dialectical law of the transformation of quantitative changes into qualitative changes, the Universe has changed qualitatively six times in the following sequence over its entire past history:

*First stage—energy evolution*
Our Universe was born and began to expand approximately 12 to 14 billion years ago from a zero point in the form of pure and weightless energy which did not contain any galaxies, stars, suns, earths, living beings, or any weighty substance whatsoever.

*Second stage—hydrogen evolution*
Primordial energy contained an encoded purposeful program which provided that the second stage of the evolutionary development of the Universe would be accompanied by the transformation of pure and weightless energy into weighty clouds of hydrogen plasma. This fiery *plasma* consisted of primary electrons and protons, which because of the extremely high temperatures could not be transformed into the simplest hydrogen atoms.

*Third stage—planetary evolution*
The extremely hot clouds of hydrogen plasma contained an

encoded purposeful program which provided for the breakaway of small "pieces" from the plasma. Because of their relatively small size, these "pieces" cooled down before the rest of the clouds, forming the planets. One such planet was our earth.

### Fourth stage—stellar evolution

The extremely hot clouds of hydrogen plasma contained an encoded reasonable program which provided for their transformation into stars. One such star was our sun.

### Fifth stage—biological evolution

The earth and the sun contained an encoded purposeful program which provided that the fifth stage of the evolution of the Universe would be accompanied by the emergence of biological cells and the appearance of living beings.

### Sixth stage—intellectual evolution

The global system of living beings contained an encoded purposeful program which provided that the sixth stage of the evolution of the Universe would be accompanied by the appearance of human intelligence.[3]

The evolutionary development of the Universe does not at all imply that weighty and visible matter are more refined than weightless and invisible ideas. For example, if at a certain stage of technological development the mind of an engineer created a computer, this does not mean that the computer is a category of a higher quality than the engineer's mind. Quite the opposite. The mind of the engineer acts as the primary (of more refined quality) creator, while the computer plays the role of the secondary (less refined!) product of the engineer's creative activity. The perfection of the human intelligence (which is weightless and invisible!) is that limit toward which a weighty and visible computer system will always strive but will never reach in the process of technological development.

By perfect analogy, if an objective idea has become incarnate in matter in the course of the evolution of the Universe, this in no way implies that matter is a category of a higher quality than the objective idea. It is exactly the opposite. In this case the objective idea serves as the primary (more refined) category, while matter is the secondary (less refined) category. The perfection of an objective idea (weightless and invisible) serves as that unattainable limit toward which unintelligent matter strives in the process of evolution but will never reach.

*Do our modern scientific theories of the evolution of the Universe confirm the atheistic principle of the "uncreatability of matter?"*

The evolutionary theory of the Universe leads to the inevitable conclusion that matter is gradually born from nothing, step by step, as the Universe expands. According to one mistaken opinion, the evolutionary expansion of the Universe could have occurred without the creation of energy.[4] Is this really true?

If matter were in fact uncreatable and indestructible, the Material World would be eternal. If the Material World were eternal and matter were uncreatable and indestructible, then according to the scientific theory of evolutionary development, all of the world's supplies of hydrogen plasma would have turned into helium a long time ago and the stars would have stopped shining.[5] But it is obvious that the heavens contain an incalculable number of shining stars. Consequently, matter is creatable, and the Material World is not eternal.

The atheistic fairy tale of the "uncreatability of matter" prohibits any quantitative changes in energy. It requires that there always be a constant amount of each type of energy (positive or negative) in the world. According to the dialectical law of the transformation of quantity into quality, the constancy of quantity would inevitably entail the constancy of quality. This means that the false principle of the "uncreatability" of matter deprives energy of its right to qualitative as well as quantitative development. Energy would never have appeared in the world under these conditions. But if it did possess some sort of magical property of "eternal existence," it would exist in the form of some sort of dead energy sea incapable of any sort of dialectical or qualitative changes and transformations.

But in reality, the theory of the evolutionary Universe convinces us that the Universe does experience qualitative changes: initially it consisted of pure and weightless energy, then this energy was transformed into hydrogen plasma, from which the stars and the planets formed, then living beings arose from inanimate substance, and now we are witnessing an irreversible process of the transformation of hydrogen into helium.

If the quantity of matter in the expanding Universe never increased, then, according to the dialectical law of the transformation of quantitative changes into qualitative changes, matter would never have been able to undergo qualitative changes. If matter never changed qualitatively, then weightless primordial energy would never have been capable of transforming into weighty substance, and inanimate substance could never have been transformed into living beings. If this were the case, then the Universe would be *non-evolutionary*. *There is no evolution without a change in quality, and there is no change in quality without a change in quantity.*

Consequently, if the expansion of the Universe from its primordial point to its current dimensions were non-evolutionary, then your and my existence right now would be impossible. But we know that we exist and that our physical bodies consist of weighty substance. This means that the Universe has expanded in an evolutionary way and that primordial positive energy at some stage in its quantitative development changed its quality and was partially transformed into substance. Consequently, *energy and substance are creatable and destructible.*

*If matter were indestructible and its quantity in the world never changed, then its quality would also never change. But the qualitative change of matter is obvious. Consequently, the quantity of matter in the world is constantly changing, i.e., matter is creatable and destructible.*

Hence, even from the standpoint of evolutionary theory, a necessary condition for the expansion of the Universe is the creation of energy from nothing. If the fact of the evolutionary development of the Universe has been definitely proven (and that is exactly what has transpired!), *then the creatability of energy and matter can be considered scientifically proven.*

The false principle of the "uncreatability" of matter has been completely discredited not only by the scientific laws of creatability but by the laws of the conservation of matter.[6] Moreover, this principle is inconsistent with both the scientific theory of the evolution of the Universe and the dialectical law of the transformation of quantitative changes into qualitative changes. Hence, if we recognize the laws of dialectics or the scientific theory of the evolutionary Universe, then we must equally recognize the law of the creatability of matter from nothing.

*Is evolution a spontaneous or a chance development?*

The natural sciences have definitely demonstrated that the evolutionary development of the Universe occurred without any sort of arbitrariness, without any sort of accidents, unambiguously, purposefully, and expediently in complete harmony with all the laws of nature. The combination of all these ideal laws of physical nature constitutes a united (single, common and coherent) program for the origin, expansion, and evolutionary development of the Universe. According to these laws and this program, *the evolutionary development of the Universe must occur in a purposeful and definite manner, in exactly this way, only this way, and no other way.*

Hence evolution is by no means a chance or a spontaneous development. Evolution is first and foremost the purposeful and programmed development of quantity and quality which occurs in stages according to certain laws. Programmed development would be impossible without a proper program. The unwritten combination of the semantic content of all the laws of nature is also a universal ideal program for material development.

The Universe could not contain the program for its own genesis, because the program of the genesis of the Universe is the cause while the genesis itself is the effect, and the cause must always come before the effect. This means that it was not the Universe which generated the program and laws for the evolutionary development of the Universe (as the atheists would like to portray it!), it was an ideal (nonmaterial, otherworldly!) program and law of development which generated matter and the material world.

The program of the genesis of the Universe could not be encoded in the Universe itself, because the Universe did not exist when it had not been born yet. Hence the combination of all the laws of nature, which constitutes a coherent (united) program for the origin, expansion, and evolutionary development of the physical Universe, is an ideal (not a material!) category. This ideal program for the material development of the Universe was embedded in that primordial ideal point with zero physical dimensions at which the Universe originated and began to expand 12 to 14 billion terrestrial years ago.

Many billions of years ago, at that unique instant, when the Universe was already destined to be born but had not yet been born, the "Universe" was an ideal point which contained no weighty substance or physical energy. Its volume, mass, and all of its material attributes were equal to ideal zero. This ideal point contained all the laws of the future nature, whose combination constituted a brilliant program for the birth and colossal evolutionary development of the Universe. This means that the semantic content of the laws of nature and the ideal program existed and still exists prior to and outside of the Material World.

The ideal program for the genesis of the Material World is independent of matter and cannot contain anything material, because prior to the birth of the Material World there was no matter, nor could there be. Hence it would be illogical to suppose that the ideal program for the origin of the physical Universe was embedded in the Universe itself. Moreover, this ideal program of material development could not arise spontaneously, by itself, without a programmer. Laws do not exist without a lawmaker, and a useful program does not exist without an intelligent programmer or Creator.[7] To suppose that the extraordinarily complicated and efficient program for the genesis and development of the Material World could come into being by pure chance or spontaneously is to believe blindly in unscientific fairy tales and descend to atheist superstition. Consequently, the ideal program for the genesis and stage-by-stage development of the Material World was created ("written") by an ideal and absolutely perfect intelligence with no material attributes. We call this Intelligent Creator of all the laws of nature and the ideal program for the material development of the Universe the *Absolute God.*

To say that the Universe originated by its own program or as a result of the operation of the laws of nature without any programmer, without any lawmaker, or without any Creator is tantamount to saying that humans are born because they want to be as a result of their own actions without any father, any mother, or any parents.

So this is a purely scientific model of the origin and development of our Universe. Does it contradict the Bible? Of course not!

According to the Bible, God's creative efforts had nothing in common with the improbable tricks of a sorcerer who instantaneously, in a fraction of a second, waved his magic wand and created the heavens, the stars, the planets, the earth, people, animals, and so forth. This is how the atheists characterize the Bible's description of the creation of the world for the purpose of making their war against religion easier. These same atheists imagined God as a decrepit old man with a long beard. Several of religion's more unsophisticated adherents, who reject science, have supported the atheists' conception of the creation of the world and have thus added a lot of fuel to the atheists' small fire, even though they are formally in opposition to them.

But in reality the Bible portrays God as an ideal, absolutely perfect category, not in the material guise of a bearded old man. This is evident from the very first verse of the Bible, which states that "in the beginning, God created the heaven and the earth." Before the creation of the heaven and the earth there was still no matter nor even "light and dark," i.e., even energy (positive or negative). In this context *matter* means everything that possesses volume, weight, or at least physical energy. According to the Bible, God (as the creator of matter!) existed prior to the creation of matter and independently of matter and consequently is an ideal category rather than a material one. Physical bodies and biological organs are not inherent attributes of the Ideal God but are products of God's creative activity.

Nowhere in the Bible is it written that God created the Material World instantly by waving a magic wand. You will find no evidence in the Bible whatsoever that the creation of the world is inconsistent with the evolutionary development of organic and inorganic matter. It is exactly the opposite. According to the Bible, the cycle of creation of the Material World by an Ideal God is gradual and purposeful (programmed) in nature. According to the Bible, the Universe was not created all at once, in one instant, but gradually, in six stages (in six Biblical days, not terrestrial days!).

The ancient Hebrew language had no special words for such concepts as, for example, "stage," "period," or "epoch." Hence the word *yom* meant not just a day but any interval of time. The six days of creation mentioned in the Bible were not 24-hour terrestrial days but stages or epochs in the evolutionary development of the Universe. Evidence of this is provided by the Bible itself, which states that the earth was created only on the "third day." This means that according to the Bible, one could speak of "terrestrial" days only after the third "day" or stage of the creation of the world.

According to the special theory of relativity, time is a function of speed. The greater the speed at which an object is traveling in relation to us, the less its time is by comparison to our time. If any elementary particle is traveling at a speed close to the speed of light relative to the earth, billions of years would elapse on the earth during a few "days" of the elementary particle's life. The primordial Universe was a plasma of pure energy whose elementary particles were traveling at speeds close to the speed of light. Hence billions of years on the earth are equivalent to one day in the life of the primordial Universe. Calculations reveal that one Biblical day is equal to approximately 2 billion terrestrial years.

A scientific analysis of the history of the Universe demonstrates that matter as a whole has exhibited a trend toward lower velocities and greater density over time. Light particles travel faster than dense systems. Hence each preceding "day" (stage) in the creation of the world was longer than each subsequent "day" (stage).

Criticizing the ancient Bible for the fact that it calls the stages in the evolutionary development of the Universe "days" is exactly the same as calling the modern English language unscientific for sometimes using the word "day" instead of the word "era." If you say that "a day of universal happiness will arrive" (instead of "an era of universal happiness will arrive"), no one would even think of criticizing you for it. Nevertheless, atheists have raised an incredible antireligious hue and cry over the Biblical "days" of creation and have tried to make them equivalent to terrestrial days.

If the Bible uses the word "day" to refer to an entire era in the development of the world, it uses the words "evening" and "morning" to express the "beginning" and "end," i.e., the starting points and end results of the era.[8] For example, the condensed Biblical expression "And the evening and the morning were the first day" should be translated into contemporary scientific language as follows: "And vacuum space (negative energy) formed and energetic photons (positive energy) formed: these were the results of the first stage in the creation of the Universe by the absolutely perfect God (energy evolution)."

Thus, in essence, the six Biblical "days" of creation of the world are the six stages in the evolutionary development of the Universe. If we say that God created the world in six Biblical days, then this means that the Universe was created in stages, in six creative stages, gradually, in an *evolutionary way*, and not instantaneously, in leaps and bounds, or in a *revolutionary way*, as the atheists and materialists who call for revolutionary upheavals claim. Hence *the scientific theory of the evolutionary development of the Universe confirms rather than discredits the Bible. Evolutionary theory is not opposed to the Bible, but to atheism*, because it leaves absolutely no room for materialistic appeals for "revolutionary leaps."

That is why the scientific theory of the evolutionary (and not steady-state) Universe is purely religious. It is the steady-state (and not evolutionary!) theory which is purely atheistic. Totalitarian atheism has abused the term "evolution" and has portrayed everything upside down, and millions of simple and naive people (religious and nonreligious) continue to blindly believe and repeat it without comprehending the meaning of the word "evolution."

The Bible has opposition not in the true scientific theory of the evolutionary development of matter, which proceeds on the basis of the Absolute God's ideal program, but in its unscientific atheist counterfeit, which we will call *pseudo-evolutionary*. Without any proof whatsoever, the pseudo-evolutionary theory proceeds *a priori* from the obviously false assumption that the evolutionary development of matter occurs spontaneously, on its own, without any ideal program and without any intelligent Creator.

The development of the Universe is called evolutionary not because it occurs on its own but because the Universe has never changed in a revolutionary manner, "in leaps," throughout its entire history. Instead it has expanded in such a way that gradual quantitative changes in the dimensions of the Universe have been accompanied by qualitative changes in six successive stages.[9]

In modern scientific terms, the qualitative changes in the Universe which have occurred in six stages indicate that in terms of their content the principles of the Bible are *dialectic and evolutionary*, even though they do not contain the modern words "evolution" or "dialectics" and to some people they might superficially seem metaphysical or even dogmatic. The Biblical principle of the creation of the world by God in and of itself contains dialectical elements, because it postulates a qualitative transformation, i.e., the evolutionary development of nonexistence into existence. And conversely, materialism, which has formally declared itself dialectical, and atheism which relies on this materialism, are in fact completely incompatible with dialectics. This is obvious from the fact that for decades atheism and "dialectical" materialism waged a merciless war against the theory of the expanding and evolutionary Universe until the obvious scientific facts forced them to surrender.

On the basis of the latest advances in the natural and philosophical sciences, we may summarize the *scientific theory of the evolutionary development of the Universe* as follows:

*1. The expansion of the Universe was accompanied by a proportional increase in the amount of matter in it. According to the dialectical law of the transformation of quantitative changes into qualitative changes, the gradual increase in the amount of matter in the entire Universe was periodically accompanied by radical qualitative changes in the Universe in six evolutionary stages. Each stage resulted in a qualitatively new form of physical existence, namely: 1) pure energy; 2) hydrogen plasma; 3) the planets; 4) the stars; 5) living cells; 6) human beings. These six forms of existence correspond to the six stages of the evolutionary development of the Universe and the six Biblical "days" of the creation of the world.*

2. The primordial zero point (at the first white cosmic hole) contained all of the ideal laws of future nature, which in combination constitute a coherent program for the origin, expansion, and evolutionary development of the Universe. But laws are not written without a lawmaker, and a program cannot exist without a programmer. We call the Intelligent Creator of all the laws of nature and the ideal program for the material development of the Universe the Absolute God.

God's absolutely perfect intelligence created the ideal laws of physical nature and the universal program for the evolutionary development of matter at a time when our Universe did not exist yet.

3. The Universe did not originate by accident or all by itself. It was created by the Absolute God according to a predetermined program. It was not created all at once or instantly, but gradually, in six evolutionary stages, each of which was qualitatively different from the other.

God created and developed matter not by means of a magic wand or with material hands, because the ideal God does not contain anything material whatsoever. God created and perfected matter and the Material World solely by means of absolutely perfect ideal Intelligence, in which any form of matter is controlled through the medium of signals and information communications. Evolutionary development did not occur by chance or by itself but in accordance with the programs and laws which were created by God and are under God's exclusive control.

4. The theory of the evolutionary development of the Universe in no way discredits the Bible. It confirms it and makes it more persuasive and less vulnerable. The Bible is not opposed by the objective science of the evolutionary development of the Universe, which follows God's laws and program. The Bible is opposed by the unscientific atheistic interpretation of evolutionary theory which states that the Universe evolves spontaneously, by itself, without any program, without any idea, without the involvement on God's part, and so forth.

So how can atheism counter the scientific theory of the evolution of the Universe, except by using its own whims and illusions? It can only counter scientific truth with its completely unsubstantiated initial assumptions that the Universe was born, developed, and evolved by chance or on its own. But then we might logically ask: where are the scientific proofs for atheism? The answer is simple: scientific atheism has absolutely no scientific proofs! It only has its convenient *initial assumptions,* which hundreds of millions of people have been *compelled to believe blindly,* even though it is logically impossible to believe in them. In fact, atheism's fairy tales about spontaneous creation, spontaneous motion, and spontaneous evolution of matter are much more unscientific and implausible than any run-of-the-mill fairy tale.

Even small children laugh gently and loudly at the hammers, buckets, rugs, chairs, and tables which move under their own power in the fairy tales, while hundreds of millions of adults have been compelled to believe blindly in the atheistic fairy tale not only of self-propelled matter but of its spontaneous evolution, which supposedly occurs without anyone's will whatsoever. Even the movement of a hammer in a fairy tale is willed by a sorcerer. But no decent fairy tale would ever try to tell you that an inanimate hammer created itself, developed by itself, and transformed itself into a live eagle without some sort of magical power.

But atheism allows itself to tell fairy tales to people in which inanimate matter creates itself from nothing, develops purposefully on its own, and transforms itself into living beings. And all of this occurs without any sorcerer, any creator, any will, and any intelligence. If the atheists were right, watches would run themselves without any springs, electronic components would be created out of nothing, these components would combine spontaneously to form televisions right at our tables, houses would build themselves without any construction workers, the glass of water we are getting ready to drink would transform itself into a roasted chicken, and so forth. But we know that these atheistic miracles have never occurred anywhere at any time.

Hence we are increasingly convinced that humanity could not come up with a greater folly than the fairy tale about the spontaneous creation and purposeful spontaneous development of unintelligent matter or the atheist superstition concerning the possibility of the spontaneous transformation of inanimate and primitive matter into the animate and highly organized matter of the human brain.

In light of all the above, we will proceed to examine and compare the scientific and Biblical models of the six stages of the evolutionary development of the Universe.

Moses makes God's word known to the people
(by Gustav Doré)

# Part Two
# Six
# Biblical
# Days

The "First Day"

Many billions of years ago there was nothing physical, neither planets, nor the earth, nor the stars, nor the sun, nor light, nor darkness, nor galaxies, nor any Universe at all. And God created the ideal program of the evolution energy. (And God said, "Let there be light.") In accordance with this program, the Universe was born 12-14 billion years ago and began to expand from nothing (from a "white hole") in the form of zero sum of positive and negative energy ("light" and "darkness.")

# =10=

## THE FIRST STAGE OF CREATION
## ENERGY EVOLUTION: LIGHT AND DARK

I n the beginning God created the heaven and the earth. And the earth
was without form, and void, and darkness was upon the face of the deep.
And the Spirit of God moved upon the face of the waters. And God said, Let
there be light: and there was light. And God saw the light, that it was good:
and God divided the light from the darkness. And God called the light Day,
and the darkness God called Night. And there was evening and there was
morning, first day.[1]

**The Biblical Model in Modern Popular Scientific Language**

1. In the beginning God created space, time, and everything which
moves and develops in space and time. "In the beginning God created
the heaven and the earth."

2. According to the ideal program of material development created
by God, the Universe originated from nothing as a zero sum of positive
and negative energy (the first stage of evolution, i.e., the "first day" of
creation). But the original positive energy had not yet assumed the
substantial form of weighty physical bodies. It consisted of an intact and
weightless continuum of formless photons whose physical volume,
weight, and rest mass were equal to ideal zero—"The earth," i.e. matter,
"was without form, and void."

The negative energy gave rise to vacuum (i.e., physical!) space,
which constitutes an unbroken continuum of antiphotons. This physical
space expanded in an ideal void—"And darkness was upon the face of
the deep."

The ideal (nonmaterial) program for the evolutionary development
of matter outlined by God was already encoded in the foundations of

this primordial Universe. According to this program, in the second phase of the evolutionary development of matter, the pure and weightless energy of the primordial Universe would subsequently be transformed into weighty clouds of hydrogen plasma—"And the Spirit of God moved upon the face of the waters."

3. And God created an ideal program of energy evolution. This program constituted a combination of laws for the purely energetic world such as the law of the creation and conservation of energy—"And God said, Let there be light." And the positive energy of photons was born from nothing—"And there was light." According to the law of conservation created by God, this event was simultaneously accompanied by the generation of the exact same amount of negative energy, from which vacuum space expanding from zero was formed.

4. Thus God split an ideal zero into a zero sum of material opposites, namely positive and negative energy—"And God divided the light from the darkness."

And God was satisfied with the results of this creation, because it included weightless and formless photons which would be transformed into weighty substance with volume—"And God saw the light, that it was good."

5. And God called the positive energy of the photons light ("day") and the negative energy of vacuum space dark ("night"). And both expanding space ("and there was evening") and energetic photon ("and there was morning") were born. Thus, the first stage of the creation of the Universe by the absolutely perfect God was completed. We call this period of creation the stage of *energy evolution*—"the first day."

## The Modern Scientific Model

The truth of the expansion of the Universe is now acknowledged unconditionally even by totalitarian atheism, although it has unjustifiably suggested that this expansion supposedly began not from an idea zero but from a superdense "cosmic egg" in which the mass of all of today's galaxies and stars was concentrated.[2]  Subsequently it was demonstrated that the expansion of this sort of "cosmic egg" would be totally impossible.[3]

At that unique instant when the Universe was destined to be born but had not been born yet, it contained nothing material whatsoever. Its volume, mass, energy, and all its material attributes were equal to ideal zero. Even the time which indicated the age of the Universe at that unique instant was exactly equal to zero for any observer (imaginary or real): $t = t_0 = 0$. If we substitute these values into the appropriate formula of the special theory of relativity we find that the newborn matter must have been traveling at the speed of light: $c = 299,792$ km/sec. Only elementary particles with a rest mass of zero, such as photons, can travel at these velocities. From this we may conclude that the Universe was born from nothing in the form of energy. In the apt words of Academician Y. B. Zeldovich, the Universe in the earliest period of its existence was a "laboratory of high energies and high temperatures."[4]

If according to the theory of relativity the Universe had to have originated in the form of pure energy, then according to the laws of conservation, this energy must have originated as equivalent opposites, namely "light" and "dark." To put it in simpler terms, the slightest (as small as possible) positive energy of a photon could not have been created without the simultaneous creation of an equally small equivalent amount of negative energy. Hence 12-14 billion years ago our Universe originated from nothing, from an ideal zero, as a zero sum of positive and negative energy, photons and antiphotons. But then we might logically ask what exactly is negative energy and in what form does it exist.

The answer to this question was provided back in 1928 by the British physicist Paul Dirac, who advanced the bold hypothesis that the "perfect vacuum" is filled to the brim with negative energy. This hypothesis was experimentally confirmed in 1932 and subsequently became a generally accepted theory which depicts the vacuum of space as a kind of "universal energy ocean." Figuratively speaking, *a space vacuum is a raging ocean of negative energy*. Vacuum space consists o negative energy just as the Atlantic Ocean consists of water.

In order to resolve the debate between atheism and religion correctly and objectively, it is first of all very important for us to understand that the space of the Material World is not a void at all and that it is the same kind of physical category as all the physical bodies we see around us every day. If space vacuum were a void, then physica bodies would not be able to attract one another in this void as required by the law of universal gravitation, which states that even in a "perfec vacuum" electrons will be attracted by an atomic nucleus, the moon will be attracted by the earth, the earth will be attracted by the sun, and the galaxies will be attracted to one another. Physical space possesses its own volume, its own boundaries, and its own energy just as the earth, for example, where we live. The only difference is that the earth consists of "clusters" or "concentrates" of positive energy and therefore has a rest mass. On the other hand, *physical space is an unbroken continuum of dispersed negative energy which has no rest mass and no weight.*

We can evaluate the degree of concentration of energy in substance and the degree of its dispersion in physical space by the following figures: one billion cubic kilometers of vacuum space contains just as much negative energy as the positive energy concentrated in millionth of a gram of substance. Each kilowatt hour of negative energy account for about 20 million cubic kilometers of space vacuum. Each cubic kilometer of space vacuum consists of $5 \cdot 10^{-8}$ kilowatt hours of negative energy.

If positive energy is the basic material and "building blocks" for weighty substance, negative energy is the "mortar" of weightles physical space. Physical space exists within certain boundaries just like any other physical object. Hence if we are not surprised by the fact that there is no earth outside the boundaries of the earth on which we live, we should also not be surprised that there is no physical space outside the Material World. The boundaries which separate the existence of physical space from its nonexistence are also the boundaries of the Material World.

Vacuum space cannot be touched with the hands or put on scales. Nevertheless it is material and not ideal, as long as we have decided to call energy a material category. Physical space is material because it is an ocean of negative energy. The Bible uses the word "dark" to refer to the negative energy of vacuum space, which consists of antiphotons. We also know that the volume, weight, and all the dimensions of photons are equal to zero. Nevertheless, they are material, and not ideal. Photons are material because they are portions of positive energy. The Bible uses the word "light" for the positive energy of photons. The Biblical expression "and God divided the light from the darkness" should be translated into modern scientific language as follows: "And God divided the ideal 'nothing' into a zero sum of energetic opposites, whose mutual annihilation is impossible under certain conditions."

At first glance a believer might think that the transition from the concept of "empty vacuum" to the concept of a "physical ocean of negative energy" would be damaging to religion, because it would leave no room for the invisible God to live. But in fact this belief is fundamentally wrong for two reasons. First of all, God is an absolute category, not a material one. This means that God exists in absolute eternity outside of any matter, outside of any time or space, and not in physical space. Secondly, the concept of an "empty vacuum" allows atheism to pass off its implausible fairy tale of the "uncreatability of matter," which has led millions of innocent people away from religion, as genuine. Indeed, if vacuum were a void, there would be no negative energy. If there were no negative energy, positive energy would be uncreatable and indestructible. But in reality everything is exactly the opposite.

Now we might ask whether modern astronomy knows of any scientific facts of the creation or destruction of matter in the Universe. A convincing answer to this question is provided by scientific theory, which has definitely confirmed the existence of a large number of so-called white and black holes in the Universe.[5] A *black cosmic hole* is the name given to an ideal point in physical space at which the mutual annihilation (disappearance) of equal amounts of the positive energy of a collapsing (dying) star and the negative energy of surrounding space concludes. The term black hole was coined in 1969 by the American scientist John Wheeler. A *white cosmic hole* is the name for an ideal point at which equal amounts of positive and negative energy are generated from nothing. It was from this sort of ideal point that the Universe was born and began to expand as a zero sum of energy opposites.

Until recently, the theory of white and black cosmic holes had not been experimentally verified. But quite recently, "quasars, which are essentially flowing collections of white holes, were detected on the periphery of the present Universe,"[6] and in late 1992 the *New York Times* published a photograph of a black hole taken at the moment of its formation by NASA.

Thus, modern science has become fully involved in studying the moment of the birth of the Universe, namely the moment when our Universe was a nonmaterial point with an ideal program of colossal development. At this primordial moment in time, the mass, energy, space, all the dimensions, and all the material attributes of this Universe were equal to zero.

That is when the first antiphoton appeared and elementary space with an unimaginably small amount of negative energy formed. This process was simultaneously accompanied by the appearance of the first photon with an equally unimaginably small amount of positive energy. All it took for the nonexistence of the Universe to become unstable was the generation of an infinitesimal volume of space vacuum.

And in fact an infinitesimal volume of space vacuum with no positive energy has a density equal to zero. In this case the negative energy of the vacuum is not balanced by any positive energy. The disturbance of this equilibrium at zero density caused a white hole to open up and a flux of positive energy to burst into this infinitesimal volume, accompanied by an equivalent flux of negative energy which expanded the boundaries of physical space.

The positive energy of photons was continuously increasing with a simultaneous and equivalent increase in the negative energy of antiphotons. As the amount of negative energy increased, the dimensions of space increased in all directions at the speed of light. Photons were dispersed in this expanding space in all directions also at the speed of light: $c = 299,792$ km/sec. Any other velocity would have been completely inadmissible for pure energy. Hence the primordial (purely energetic) Universe could not have increased its rate of expansion gradually from zero to light speed. It bypassed this range of velocities completely and began its expansion at the speed of light from the very beginning, immediately, instantaneously. The generation of a colossal amount of pure and weightless energy from the first white hole, which began suddenly and immediately at that colossal speed known as the speed of light has come to be known as the *Big Bang*.

No matter how the arithmetic quantities of positive and negative energy may have increased individually, their algebraic sum has always been and is still constant and equal to zero in full accord with the law of conservation of energy. This means that the primordial Universe consisted of the negative energy of vacuum space and the positive energy of photons. But the positive energy of photons ("light") was separated from the negative energy of vacuum space ("dark") in the sense that their mutual annihilation (coalescence or disappearance) became impossible until the onset of a so-called *gravitational collapse* accompanied by the formation of black cosmic holes, at which the positive energy of photons and the negative energy of vacuum space will disappear together after tens of billions of years. Thus, the Universe was born and began to expand 12 to 14 billion terrestrial years ago as the zero sum of continuously increasing energy opposites, namely positive and negative energy, photons and antiphotons, i.e., light and dark. This definitely implies that there is a nonphysical medium in which our Universe originated and is expanding. In contrast to physical space, we will call this qualitatively different medium, which contains no physical components (even energy!), *ideal space.*

The continuous generation of a zero sum of positive and negative energy from nothing is the first step in the qualitative transformation of nonexistence into the existence of the Universe. That is why we call it the stage of *energy evolution* (the "first day"). But while at the first stage of the Universe's evolutionary development an objective idea was embodied in a zero sum of positive and negative energy, this in no way implies that physical energy is of a supposedly higher quality than the objective idea. In this case, the objective idea acts as the project manager, and physical energy is only the basic raw material which faithfully carries out the orders of the objective idea in the evolutionary development of the Universe.

But nothing material can be born by itself without an external cause. A child is not born without a mother, an egg is not hatched without a hen, an apple does not grow without an apple tree, and substance is not generated without energy. But what kind of external force generated the zero sum of positive and negative energy? Whose brilliant mind divided nothing into a zero sum of real energy opposites? The only thing that is clear is the fact that a colossal (but not physical!) force which compelled the Universe to be born and develop in a certain way was already present at the ideal point of the first white cosmic hole. Atheists call this force the laws of nature, while Shklovsky calls it a "super-gene with an incredible array of potentials."[7] We call it the ideal program for the genesis and development of the Material World. We call it ideal because it was embedded in the ideal point of the white cosmic hole before the Universe was born. We call it a program because the semantic (ideal) content of the laws of nature is essentially a program which determined the origins and rules for the subsequent behavior of the Universe.

But names will not change the essence of the question, because any law of nature is an individual rule in the universal ideal program created by God and encoded at the energy level. The ideal program of material development is a complete collection of all the laws of nature created by God. While a law of nature defines the rules of behavior for particular material elements and systems under specific conditions, the ideal program of material development unambiguously controls the behavior of the entire Universe as a whole. In any case pragmatic laws would be impossible without an intelligent lawmaker, and a program of development would be impossible without an intelligent and provident programmer. The primary source of all that exists is that absolutely perfect intelligence whose laws and whose program determined the genesis and development of the Universe. We call this intelligence the ideal (nonmaterial) God. We call God nonmaterial because, as the creator of the Universe, God is outside the Universe, outside the Material World, and outside matter altogether.

On the basis of the latest advances in modern science, we may summarize the creation and first (energetic) stage in the evolutionary development of the Universe as follows:

*1. The Universe in which we now live was originally an ideal point, and all its material attributes were equal to ideal zero. It contained nothing physical, not even pure and weightless energy. This zero point contained the ideal laws of the future nature. The combination of all these future natural laws was a brilliant program for the birth and colossal evolutionary development of the Universe in the future. This ideal program of material development could not have arisen spontaneously by itself. It could only have been "written" by an ideal (nonmaterial) and absolutely perfect intelligence, which we call God.*

The primary category is the Absolute God, who exists in absolute eternity outside any space and time. The secondary category is the product of the Absolute God's creative efforts, namely the entire Relative World, i.e. space, time, and everything which moves and develops in space and time ("In the beginning God created the heaven and the earth").

2. According to the ideal program of material development, the Universe was born and began to expand approximately 12 to 14 billion terrestrial years ago from a zero point, from ideal zero, from nothing, as a zero sum of positive and negative energy in full accord with the laws of conservation. This ideal point, from which pure physical energy was born as a powerful gushing sphere, is called the first cosmic white hole. That is how the so-called "Big Bang" occurred. The newborn Universe was a purely energetic system and did not contain any galaxies, any stars, any planets, any physical bodies, any weighty substance, any molecules, any atoms, neutrons, protons, electrons, or positrons whatsoever. It was a coherent and weightless continuum of formless photons whose physical volume, weight, and rest mass were equal to ideal zero (The Universe "was without form and void").

The ideal (nonmaterial) program for the evolutionary development of matter created by God was encoded in the newborn Universe at the photon (energy) level. According to this program, in the second stage of the evolutionary development of matter, the pure and weightless energy of the primordial Universe would subsequently be transformed into weighty clouds of hydrogen plasma ("And the Spirit of God moved upon the face of the waters"). Here the word "water" expresses the concept "hydrogen plasma."

3. The continuously generated negative energy ("the dark") took the form of a vacuum (physical) space, which is continuously expanding in an ideal void. This means that the three-dimensional physical space of our Universe was born and has been expanding in a multidimensional ideal space as a raging ocean of negative energy ("and darkness was upon the face of the deep"). All of the boundaries of this ocean have been continuously receding from its center at the speed of light: $c = 299,792$ km/sec. At the same time

*the continuously generated positive energy ("light") took the form of photons, which are flying in all directions at the speed of light within the boundaries of expanding vacuum space ("And God said, Let there be light, and there was light").*

*4. Thus, God divided an absolute and ideal zero into a zero sum of material opposites, namely positive and negative energy. God separated positive energy from negative energy so that they would not coalesce and undergo mutual annihilation for many billions of years ("and God divided the light from the darkness").*

*And God was satisfied with the results of the work of creation, because they contained weightless and formless photons which would subsequently be transformed into weighty substance with volume and rest mass ("And God saw the light, that it was good").*

*5. The positive energy of the dispersing photons came to be known as "light," while the negative energy of expanding space came to be known as "darkness." And thus expanding space arose ("and there was evening"), along with energetic photons ("and there was morning"). Thus went the first stage in the creation of the Universe by the absolutely perfect God. We call this period of creation the stage of energy evolution ("the first day").*

The above is a purely scientific model of the birth and first stage of the evolutionary development of the Universe. So how does it differ from the Biblical model? By analysis and comparison we can see that the scientific and Biblical models are not essentially different and have the same semantic contents. The only difference lies in their form and style. For example, the ancient Hebrew language of that time did not yet include the concept of "energy." Hence the Bible uses the word "light" to express the concept of positive energy and the word "dark" to express the concept of negative energy. The Biblical model of the first day of creation of the world is completely consistent with the scientific theory of energy evolution, which states that energy opposites were constructed from nothing and were then separated from each other.

The scientific model confirms rather than discredits the Biblical model. The first stage of the expansion of the Universe was accompanied by the continuous and simultaneous generation of equal amounts of positive and negative energy ("light" and "dark"), in full accord with the laws of conservation and creation of energy.

It is the scientific theory of the *evolutionary* (not the steady-state) Universe that informs us of the first stage of energy evolution, i.e. of the *qualitative transformation* of nonexistence into existence of an energetic Universe, thus confirming the Biblical model of the creation of a zero sum of energy opposites, i.e., "light" and "dark," from nothing. It is the "theory" of the steady-state (and not the evolutionary) Universe which depicts the latter as eternal and unchanged and thus tries to refute the Biblical model of the creation of the world.

The end of the stage of energy evolution merely signified the beginning of the stage of hydrogen evolution. But it in no way means that energy evolution somehow stopped. The simultaneous production of positive and negative energy in new white cosmic holes continued in all the subsequent stages of the evolutionary development of the expanding Universe.

Under the pressure of scientific facts, even atheism has been compelled to admit that the gigantic Universe originated at a zero point. But atheism, referring to its old fable about the "uncreatability" of matter, has tried to "squeeze" an infinitely large mass into this original point. To say that matter is uncreatable and at the same time admit the truth of the expansion of the Universe is, in the final analysis, equivalent to "squeezing" the entire Universe, including our earth, the sun, all the stars, and all the galaxies, into a single original point with zero volume. No ordinary teller of fairy tales would even think of saying something so preposterous. The only place you can hear it is in atheist "science." In fact, fairy tales limit themselves to describing large genies squeezing into small bottles. But no teller of fairy tales would permit himself or herself a fantasy such as squeezing "an infinite mass into a zero point." Atheism has not only allowed itself to tell such falsehoods to people but has even had the gall to call them "scientific." And this has been done at a time when religion is modestly referring to its scientific truths as "faith."

Atheism is "scientific" only in formal terms, on its cover, on its label, and in its name. But in essence and in terms of its entire contents, atheism is an unscientific superstition because it is based on a blind faith in such falsehoods as "the infinite density of a zero volume," a purposeful program of development without an intelligent programmer, natural laws without a lawmaker, a product of creativity without a creator, and so forth.

But if so, why is atheism shouting so loudly that science supposedly discredits the Bible? It is shouting because it cannot do anything else. Loud and raucous shouts, effective and constant propaganda, and scientific misinformation for the masses more than compensate for the lack of scientific facts in atheist beliefs. Yemelyan Yaroslavsky, the founder of Soviet atheism, wrote that "The Biblical legend that God somehow created light before the sun appeared is an absurdity and fabrication of a savage, benighted, ignorant man."[8] But modern science has established that light existed in the form of photons long before the stars and the sun were formed. Even materialists now admit that "light accounted for most of the physical matter in the Universe in the early stages of its expansion."[9]

From the dialectical law of the negation of the negation, which Marxism has taken for its own, it definitely follows that the sun radiates light energy because it itself was formed at one time from the energy of primordial light, from primordial photons. But where did these primordial photons come from? They were created and are still being created from nothing in white cosmic holes. By whom? It would be better not to ask atheists this question at all, because they would go completely out of their minds. "That's not a scientific question!" they would say. But if science is incapable of answering this basic question, then why does atheism call itself scientific? And by what right does it preach the obviously false idea that science supposedly discredits the Bible to the entire world? If atheist science is helpless against the Bible then this in no way implies that it has to convert its scientific helplessness into evil slanders against the Bible and God.

Only the following is still unclear: was it true that Y. M. Yaroslavsky really did not know the basics of modern science, or did he intentionally pretend to be a modern "savage and ignoramus" who wrote unscientific atheistic fairy tales which posited that there was no light before the sun and stars were formed?

Yaroslavsky seemed not to have been aware that it was the primordial photons (portions of light energy) which were transformed into the clouds of hydrogen plasma from which the stars and sun were subsequently formed. How could he possibly know these truths when he lived in the time of advanced science and spacecraft? It turns out that it was not Yemelyan Yaroslavsky (1878-1943), the father of modern "scientific" atheism, who knew modern science, it was the author of the ancient Bible, who lived 3300 years ago in the age of the bow and arrow. An ancient arrow which can catch up to and overtake a modern rocket is magnificent indeed.

The primordial Universe began to expand from a "point" which contained an ideal program for the genesis and development of the Universe. The general significance of this original ideal point lies in the fact that the Universe could not have been born or developed without it. Joseph Shklovsky called the newborn Universe a "super-gene" in which the entire program for its later evolution was encoded.[10] The significance of this primordial "point" in the first stage of the development of the Universe lies in the fact that without it, energy evolution would have been impossible. The initial state of the first stage of evolution was the primordial point (the first white hole), while its inevitable end result was primordial energy. But why was this primordial energy necessary?

The "Second Day" (Evening)

And God created an ideal program of hydrogen evolution, in accordance with which an expanding vacuous space ("darkness") was formed from negative energy, and positive (luminous) energy was transformed into billions of clouds of hydrogen plasma, whose accumulation is conventionally called a protogalaxy. These clouds were separated from each other by empty space. ("Let there be a firmament in the midst of the waters, and let it divide the waters from the waters.")

= 11 =

# THE SECOND STAGE OF CREATION
# HYDROGEN EVOLUTION: ENERGY AND
# SUBSTANCE

## The Biblical model in Ancient Hebrew

And God said, Let there be a firmament in the midst of the waters, and let it divide the waters from the waters. And God made the firmament, and divided the waters which were under the firmament from the waters which were above the firmament: and it was so. And God called the firmament Heaven. And there was evening and there was morning, second day.[1]

## The Biblical Model in Modern Popular Scientific Language

And God said: let the positive energy of photons be converted into clouds of hydrogen plasma ("waters") and let vacuum space ("firmament") separate each hydrogen plasma cloud from all the others. And God created the space between the clouds of hydrogen plasma both within and outside of each protogalaxy (future galaxy). And it was so. And God called the space heaven. And a pure field of vacuum space arose ("and there was evening") and clouds of hydrogen plasma arose ("and there was morning"): thus went the second stage of the creation of the Universe by the absolutely perfect God. We call this period of creation the stage of *hydrogen evolution ("second day")*.

## The Modern Scientific Model

The natural sciences have definitely established that our Universe did not yet exist many billions of years ago. At that unique instant when it still had not been born but was destined to be born, the Universe was an ideal point (the first white cosmic hole) which did not contain anything material. Its energy, mass, volume, and all its material attributes were equal to ideal zero.

This first white cosmic hole already contained an ideal program for the genesis and subsequent material development of the Universe, which basically consisted of the semantic contents of the complete combination of all future natural laws. According to this program, the Universe was born 12 to 14 billion terrestrial years ago from nothing as a zero sum of positive and negative energy in full accord with the laws of conservation. As it was generated, negative energy took the form of physical (vacuum) space, which has continuously expanded in all directions at the speed of light. At the same time the positive energy generated in the white hole took the form of photons, which are also dispersing in all directions at the speed of light within the confines of expanding physical space.

The primordial space in the environs of the white hole was filled to the brim with photons, whose weight, volume, and all dimensions were equal to zero. These photons contained the encoded program for the second stage of the evolutionary development of the Universe, in which weightless photons would be logically transformed into clouds of hydrogen plasma (i.e., "waters" in the Biblical terminology): and God created an ideal program of hydrogen evolution ("And God said, Let there be a firmament in the midst of the waters, and let it divide the waters from the waters").[2]

Figuratively speaking, the newborn Universe consisted of negative energy (expanding space) filled to the brim with an equal amount of positive energy (dispersing photons). This state of the photons in vacuum space generated an extremely high temperature, which made it inevitable for the pure and weightless positive energy to be converted into weighty electrons and positrons, followed by the formation of hydrogen plasma ions, i.e., protons and electrons.

In any white hole, all the conditions for the formation of hydrogen plasma are present. These conditions are primarily expressed in the saturation of physical space with the positive energy of photons. Despite the high saturation, these photons cannot be in a state of rest and must travel in a vacuum at an extremely high velocity approximately equal to 299,792 km/sec. This high-velocity motion of photons in a vacuum space packed to the brim with them generates an extremely high temperature and, thus, leads to the simultaneous birth of substance and electrical antisubstance such as electrons and positrons, protons and antiprotons, and so forth.

Now let us examine the mechanism for the transformation of primordial energy into substance. From physics we know that if two energetic photons collide with one another, they will be transformed into an electron-positron pair. In contrast to the photons, whose volumes, rest masses, and charges are equal to zero, the electron and positron have non-zero volumes and rest masses which are equal in magnitude but opposite in sign. They also possess electrical charges which are also equal in magnitude but opposite in sign. This was how elementary particles of substance as "clusters" or "concentrates" of positive energy formed.

An electron, which has a negative electrical charge, and a positron, which has a positive electrical charge, are attracted to one another by electrostatic forces as well as gravitational forces. Hence at low temperatures the newborn electrons and positrons will combine and be reconverted into photons immediately after they are born. The relationship between the processes of the generation and annihilation of electron-positron pairs is basically determined by the ambient temperature. As the temperature rises the generation of electron and positron pairs predominates over their annihilation. Equilibrium between generation and annihilation occurs at a temperature of approximately 10 billion °K. While the generation of electron and positron pairs predominates over their annihilation at higher temperatures, the mutual annihilation of positron-electron pairs predominates over generation at lower temperatures. While at a relatively low temperature a photon medium contains only a small number of electron-positron pairs, at high temperatures it contains a much larger number.

According to I. D. Novikov, the temperature of the newborn Universe was much higher than 10 billion °K.[3] That is why it was the site of an intense process of the transformation of photons into electron-positron pairs, i.e., a process of the transformation of pure and weightless energy into weighty substance. In the process, collisions occur not only between photons, but also between electrons and positrons. While the mutual collision of photons causes two photons to disappear and transform themselves into electrons and positrons, the mutual collision of electrons and positrons will always, at any temperature, result in the annihilation (disappearance) of only their electrical charges, without transforming them into anything else, because they are equal in magnitude and opposite in sign. But their positive rest masses are not lost all the time, only at relatively low temperatures.

If an electron collides with a positron at a sufficiently high temperature (above 10 billion °K), that is if an electron-positron pair collides in an environment saturated or supersaturated with photons, then the pair cannot lose its rest mass and be converted into a pair of weightless photons. This process results in the formation of an electrically neutral substantial particle, whose mass will be twice that of an electron. The percentage of these particles in the environment will be greater the higher the temperature is. Substantial particles of this type can be generated not just from one but from a certain number of electron-positron pairs.

But these "clusters" of substance are only stable when their rest mass is equal to $1.67495 \cdot 10^{-27}$ or $1.67265 \cdot 10^{-27}$ kg. In the first case an electrically neutral *neutron* consisting of approximately 1840 paired electrons and positrons is generated. In the second case a positively charged *proton*, in which the number of positrons is one greater than the number of electrons, is generated.

At sufficiently high temperatures (above 100 billion °K), the percentage of protons and neutrons in the primordial medium was approximately the same. A neutron has a greater mass than a proton. This means that the generation of a proton is more energetically advantageous at relatively low temperatures. Because of this, as the medium cools, a proton becomes more stable than a neutron. Hence ultimately there are more protons than neutrons in the Universe. If the temperature of the newborn Universe had dropped immediately after the formation of electron and positron pairs, then these pairs would have been annihilated (disappeared) without turning into protons or neutrons. If this were the case, then primordial energy would have never been converted into substance. But the facts say otherwise, and therefore the presence of substance convinces us that protons and neutrons were generated at a time when the first white hole had still not closed and its temperature was in excess of 10 billion °K.

All other combinations of electrons and positrons, except for protons and neutrons, are completely unstable and thus quickly disintegrate. This convinces us that the formation of electron-positron pairs is a very highly efficient property of photons, while the laws of nature were well considered and correspond fully with the requirements for the formation of stars and planets in the future. Thus, the formation of the stars and planets was programmed in advance. But who did it?

So, in any case, the number of electrons is equal to the number of protons. Each pair consisting of one proton and one electron is a split hydrogen atom. We call this kind of highly heated gas, which consists of positively and negatively charged hydrogen ions, *hydrogen plasma*. Plasma is the fourth state of matter after solids, liquids, and gases. The higher the temperature of the primordial medium was, the faster the photon plasma would have been converted into hydrogen plasma, i.e., the faster pure energy would have been converted into substance. As the primordial environment cooled down, these processes became less intense. But hydrogen plasma contains other split atoms besides hydrogen atoms. The temperature of the newborn Universe was far in excess of these levels. Therefore, the primordial hydrogen plasma contained split atoms and other chemical elements, but in small quantities. That was how the first hydrogen plasma cloud formed in our Universe.

The conversion of the pure energy of photons into the substantial mass of nuclei must have reduced the temperature of the primordial plasma significantly. But this reduction did not occur right away. After all, the white hole remained open for some time and spewed forth a spherical stream of pure energy to replace the energy which had already been converted into substance.

As long as the white hole was open, the temperature could not have fallen. But when would it close? A white hole cannot close until each newborn photon had receded from it to overtake its corresponding antiphoton, because in this situation the density of the mass near the white hole was constant: $5 \cdot 10^{-34}$ kg/cm$^3$. The white hole could not close until this density had at least doubled. Let us examine the mechanism by which this increase in density occurred.

According to the law of universal gravitation, all particles of newborn substance would be attracted to one another by gravitational forces. The center of attraction was apparently located alongside or even inside the white hole. This is why the primordial cloud of hydrogen plasma would be condensed around the white hole, thus raising the density of mass in this area. As soon as the density of the center surpassed $10^{-33}$ kg/cm$^3$, the white hole closed. In the process it turned out that the first photons had already been converted into substance and were attracted to the center, while the last photons were receding from the center at the speed of light c = 299,792 km/sec. But these photons could not reach the boundaries of physical space, which had already receded quite far and which were still receding at the same speed of light. Hence as soon as the white hole closed at the center, areas with zero density formed at the periphery of the newborn Universe and became the sites of new white holes.

When the first white hole closed, the influx of energy stopped, most of the photons were transformed into substance, while a smaller portion moved into the surrounding space. Despite the condensation, under these conditions the transformation of energy into substance caused the temperature of the rarefied cloud of hydrogen plasma to fall very slowly from $10^{15}$ to 10,000 °K. We call the central portion of this relatively "cold" cloud of hydrogen plasma which formed at the site of the old white cosmic hole a *protostar*, because these are the parts of hydrogen plasma clouds from which stars form. Cooling caused a decrease in the relative percentage of neutrons in the plasma cloud and a corresponding increase in the proton content. Preliminary estimates indicate it took about 2500 terrestrial years for the first protostar to form.

In the meantime, new white holes were developing on the periphery of the newborn Universe in the exact same way as the original white hole did. Their development also culminated in the formation of separate clouds of hydrogen plasma and protostars in the areas of old holes and a series of new white holes at the boundaries of the Universe. This process was repeated over and over again. But now not only the particles of one cloud but different hydrogen clouds were attracted to one another. They traveled toward their common center and in a circle around this center. That is why each new series of white holes opened farther and farther away from a cluster of protostars. The newborn protostars were much less and much more slowly attracted to an old cluster of protostars, some of which might already have been transformed into stars. Finally we arrive at a situation where the newborn stars formed new clusters and were no longer attracted to the old cluster of stars and protostars.

Each such cluster of protostars is commonly known as a *protogalaxy*, while a cluster of stars is known as a *galaxy*. At present there are an extraordinarily large number of galaxies. *Quasars*, which are essentially flowing chains of white holes and the hot nuclei of future galaxies, have been detected on the periphery of our Universe. The process of the expansion of the Universe and the formation of new galaxies continues.

The careful reader might logically ask why white cosmic holes generate protons, and not their electrical opposites, the antiprotons. We cannot rule out the possibility that, under these conditions, protons and electrons are stable, whereas antiprotons and antielectrons are unstable. Most scientists, however, believe that for each cloud of hydrogen plasma consisting of electrons and protons there must be one cloud of antihydrogen plasma consisting of positrons and antiprotons, even though this does not necessarily follow from the basic law of material opposites, which states that negative and positive electrical charges cannot exist without one another.

The probabilities for the formation of separate electron-proton and positron-antipositron combinations are equal. But protons and antiprotons, like positrons and electrons, have different electrical charges and like masses. Hence they are attracted to one another with a greater force and are immediately annihilated (disappear) and are converted into pure energy. This means that it would be completely impossible for them to form simultaneously in the same plasma cloud.

If a proton and electrical antiproton were to be generated in the same white hole which formed one star system, they would be immediately converted into pure energy. In that case, the Universe would not undergo any evolutionary development. The coappearance of clouds of hydrogen plasma and antihydrogen plasma within the same galaxy would also be improbable. If this were not the case, the danger of the disappearance of substance in the Universe would be extraordinarily great. The most likely scenario is that hydrogen plasma and anti-hydrogen plasma clouds are formed in different galaxies. For every galaxy consisting of substance, there must be one galaxy consisting of electrical antisubstance. Which particular galaxy consists of substance and which particular galaxy consists of antisubstance depends on the ideal program which says that either the proton or antiproton must be more stable.

Thus, in the development of some galaxies, electrons and protons are clustered in separate clouds of hydrogen plasma. In the development of other galaxies, positrons and antiprotons are clustered in separate clouds of antihydrogen plasma. The genesis of hydrogen or anti-hydrogen plasma clouds is simultaneously accompanied by the appearance of their magnetic fields. Galaxies are separated by such great distances that it would be impossible for them to meet during the expansion of the Universe. Therefore, the possibility of the meeting and mutual annihilation of hydrogen and antihydrogen clouds may also be ruled out, at least as long as the Universe is expanding.

If the original white hole were the only one in the Universe and if the energy capacity of this single hole had remained absolutely constant for any time interval, then only one indivisible hydrogen cloud would have formed in the Universe and would have been condensed until substance and antisubstance annihilated each other. But it has been scientifically proven that white cosmic holes are being born, multiplying, and dying at different edges of the Universe.

In addition, according to the laws of dialectics, the energy capacity of each white hole must be a periodic function of time and cannot remain constant. This function may be roughly approximated by a sine curve. This means that every white hole "ejects" positive energy in all directions not only in microquanta but in large "macroquantum" waves. In all probability, each gigantic half-wave forms one galaxy. While the conditions in one half of the wave are right for the formation of substance, the conditions in the other half are right for the formation of antisubstance. While the half-wave which generates substance is concentrated around the center of one galaxy, the half-wave which generates antisubstance is concentrated around the center of another galaxy. This explains why hydrogen and antihydrogen clouds in the primordial Universe became so distant from one another. The wave-like fluctuations in energy concentrations and dispersions also constituted one of the primary sources for the breakup of both hydrogen and anti-hydrogen primordial clouds.

The end of the stage of hydrogen evolution merely marks the beginning of the stage of planetary evolution. But it in no way implies that hydrogen evolution has ceased. The transformation of the pure energy of white cosmic holes into clouds of hydrogen or antihydrogen plasma has continued throughout all the subsequent stages of the evolutionary development of the expanding Universe.

On the basis of the latest advances in the natural sciences, we may summarize *the law of the transformation of pure energy into a cloud of hydrogen plasma* as follows:

*According to the ideal program for the material development of the Universe, the high-velocity motion of photons in newborn space, which was saturated with pure positive energy, generated an extremely high temperature (in excess of 10 billion degrees on the Kelvin scale) and inevitably led to the transformation of photons into electron-positron pairs and the transformation of electron-positron pairs into protons and electrons, which are split hydrogen atoms. Thus, at an extremely high temperature, the pure energy of a white cosmic hole is transformed into a hydrogen plasma in the same way that water vapor on the earth is transformed into rain clouds.*

*According to the law of universal gravitation, the cloud of hydrogen plasma is condensed by gravitational forces and rotates around its center. Condensation causes the density of the cloud to increase, and this increase in density causes the white hole to close and stop generating energy.*

*The density of the periphery of the Universe decreases as the density at its center increases. At some point in time, new zero-density sites formed at the periphery of the Universe and gave rise to new white holes. These holes develop on the periphery of the Universe in the exact same way as the first white hole did in its time. Their development also culminates in the formation of separate clouds of hydrogen plasma and a new series of white holes at the Universe's new boundaries. This process is repeated over and over again until a huge cluster of hydrogen plasma clouds, which we call a galaxy, is formed. The further expansion of the Universe has led to the genesis of more and more new series of galaxies at its periphery.*

And God created an ideal program of hydrogen evolution. *According to this program and in full accord with all of its laws, the positive energy of photons was converted into clouds of hydrogen plasma. The condensation of these clouds opened up vast expanses of vacuum space, which separated each cloud of hydrogen plasma from all the others.* ("And God said, Let there be a firmament in the midst of the waters, and let it divide the waters from the waters."[4])

So the positive energy of white holes was transformed into clouds of hydrogen or antihydrogen plasma, which are separated from one another by pure space both within and outside each galaxy. ("And God made the firmament, and divided the waters which were under the firmament from the waters which were above the firmament: and it was so."[5])

*We use the word "heaven" for the vast expanses of pure (vacuum) space which separates the clouds of hydrogen plasma from each other. And a pure field of vacuum space arose ("and there was evening") and clouds of hydrogen plasma arose ("and there was morning"): thus went the second stage of the creation of the Universe by the absolutely perfect God. We call this period of creation the stage of hydrogen evolution ("second day").*

So what does atheism have to offer to counter this scientific law of the transformation of pure energy into hydrogen plasma except for its own whims and illusions? It can only counter the scientific facts with a play of words such as the following: "there is no basis in speaking about the transformation of mass into energy or energy into mass."[6]

According to the law of the convertibility of substance and energy, *under certain conditions pure and weightless energy may acquire rest mass and be completely transformed into weighty substance.*[7] Energy is not converted into mass, but it does acquire rest mass and is thus converted into weighty substance.[8] These sorts of word games are mere gibberish, not science. Nevertheless, atheism has used them to divert millions of uninformed and trusting people from the truth.

But then we might ask where the scientific proofs of atheism are. The answer is simple: atheism has no scientific proofs and it could not have them! It only has *atheist tricks* which are designed to keep hundreds of millions of people ignorant of the theory of relativity and, thus, unable to see the difference between "mass" and "rest mass." As a result, hundreds of millions of trusting souls still believe these atheist tricks, even though it is logically impossible to believe them. If energy could not be converted into substance, then substance could not be converted into energy. If substance could not be converted into energy, then hydrogen could not be converted into helium. If hydrogen could not be converted into helium, then our sun would not give off any energy. If our sun did not give off any energy, then you and I would not be here. But we exist and develop in spite of any atheistic tricks and prejudices. In the words of Hendrik Lorentz, the Dutch physicist, even the rest mass of an electron is totally energetic in origin. And the leading scientist of the twentieth century, Albert Einstein, wrote the following on the subject: "Inert mass is nothing more than latent energy."[9]

"And God said, Let there be a firmament in the midst of the waters, and let it divide the waters from the waters."[10] The concept of "hydrogen" did not exist in the ancient Hebrew language of that time. Hence the Bible uses the word "waters" to express the concept of hydrogen. The Biblical model of the second day of the creation of the world is fully consistent with the scientific theory of hydrogen evolution, which states that positive energy yielded clouds of hydrogen plasma separated by vacuum space consisting of negative energy.

Primordial energy contained an encoded program which dictated that the second stage of the evolution of the Universe would be accompanied by the transformation of positive energy into primordial clouds of hydrogen plasma and negative energy into expanding space. The clouds of hydrogen plasma separated and became so far apart from one another that a huge but empty space was left between them. Compare this scientific concept with the Biblical expression "And God made the firmament, and divided the waters." The significance of the existence of primordial energy primarily lies in the fact that hydrogen evolution would have been impossible without it.

The initial state of the second stage of the evolution of the Universe ("the second day" of creation, which was equal to several billion terrestrial years) was energy, while the inevitable end result of this stage was space and substantial clouds of hydrogen plasma. But why were these clouds of hydrogen plasma necessary?

The "Second Day" (Morning)

According to the program of hydrogen evolution, produced by God approximately 6-10 billion years ago an enormous number of protogalaxies were formed from perfect luminous energy, which flew away from each other to great distances. Thus arose huge expanses of vacuous space within as well as outside of every protogalaxy. ("And God made the firmament, and divided the waters which were under the firmament from the waters which were above the firmament.")

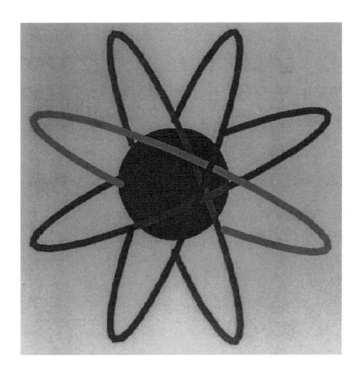

The "Third Day" (Evening)

And God created an ideal program of planetary evolution, in accordance with which deep within "furnaces" of plasma atoms of heavy elements arose from split atoms of hydrogen.

The "Third Day"

According to the program of planetary evolution, produced by God, hot clusters of the heavy elements were "shot out" from the depths of the hydrogen plasma to the periphery. They cooled off, were drawn to each other, and became the nuclei of the future planets. This is how our plant Earth was formed. ("And God said, let the dry land appear. And it was so. And God called the dry land Earth.")

## THE THIRD STAGE OF CREATION
## PLANETARY EVOLUTION: THE EARTH AND
## PLANTS

**T**he Biblical model in Ancient Hebrew

And God said, Let the waters under the heaven be gathered together unto one place, and let the dry land appear: and it was so. And God called the dry land Earth, and the gathering together of the waters God called Seas: and God saw that it was good. And God said, Let the earth bring forth grass, the herb yielding seed, and the fruit tree yielding fruit after its kind, whose seed is in itself, upon the earth: and it was so. And the earth brought forth grass, and herb yielding seed after its kind, and the tree yielding fruit, whose seed was in itself, after its kind: and God saw that it was good. And there was evening and there was morning, third day.[1]

### The Biblical Model in Modern Popular Scientific Language

And God said: let one of the clouds of hydrogen plasma in the Galaxy be condensed so that planets separate from it. And it was so. And God called one of these planets Earth, and God called those depressions on the earth into which water subsequently flowed the oceans and the seas. And God saw that the earth was suitable for life. And God said: let the earth acquire the "seeds of biological life" from which different species of plants, namely grass, the herb yielding seed, and the fruit tree yielding fruit after its kind will later grow. And it was so. And the earth subsequently produced different species of plants from those "seeds of biological life" which it had inherited from the cosmos. Each plant was born from a seed, developed, matured, yielded seed after its own kind, aged, died, and was then reborn from a seed. And God was satisfied with the results of his creation. And the earth formed ("and there was evening") and plants appeared on it ("and there was morning"): so went the third stage of the creation of the Universe by the absolutely perfect God. We call this period of creation the stage of *planetary evolution* ("third day").

## The Modern Scientific Model

There are very many scientific models of planetary evolution. Figuratively speaking, there is a different model of planetary evolution for each astronomer in the world. But in general, modern science supports the model of Immanuel Kant (1724-1804), the German philosopher and scientist, and Pierre Laplace (1749-1827), the French mathematician, which was later developed and elaborated by Alfven, the Swedish physicist and astronomer, and a number of other scientists.[2] Here we will provide a brief synopsis of this scientific model.

In the continuously expanding primordial vacuum space, pure positive energy was transformed into huge clouds of plasma which mostly consisted of hydrogen (or antihydrogen) with small admixtures of other chemical elements. One of the clouds contained an encoded program for its future evolution, which would inevitably transform it into the solar system. But how was this program expressed specifically and carried out in practice? It was primarily expressed in the law of universal gravitation and the law of conservation of angular momentum. The law of universal gravitation tells all the substantial particles of one and the same cloud to stay in place and not disperse into the surrounding space. Hence all substantial particles which cannot break away from their clouds must travel in two directions: in a circular orbit and on a radius toward the center. As a result of this particle motion the entire hydrogen cloud as a whole was compelled to undergo simultaneous condensation and rotation about its center.

Roughly 8 billion years ago this slowly rotating cloud of hydrogen plasma was extremely rarefied and had a very high angular momentum, meaning that its original dimensions were huge, on the order of several light years ($10^{13}$ to $10^{14}$ kilometers). The cloud had a magnetic field, and its temperature did not exceed 4000 to 5000 °K. In order to conserve total angular momentum, the product of the moment of inertia by angular velocity must always be a constant number. As the cloud condensed, its moment of inertia decreased. Consequently, the hydrogen plasma cloud had to increase its rotational velocity during condensation or condense with an increase in its rotational velocity. Hence this isolated cloud of hydrogen plasma condensing as a result of gravitational forces rotated faster and faster.

As condensation progressed, the magnetic field grew stronger, and the temperature and pressure within the cloud increased. But the temperature of the plasma decreased from the center to the periphery because the outside layers of the plasma cloud gave off their heat to the surrounding space.

In the depths of this still not very hot cloud of hydrogen plasma, local (comparatively small) active regions with extremely high temperatures developed for some reason or another. If the temperature in one of these regions reached $10^7$ °K, then the hydrogen in this region would burn, yielding helium and releasing heat. Helium burns at $10^8$ °K, yielding carbon and oxygen. Carbon burns at a temperature of 500 million degrees on the Kelvin scale, yielding magnesium and sodium. And at temperatures above one billion °K, oxygen burns yielding sulphur, phosphorus, silicon and so forth.

The heavier elements form in a different way. If the ambient temperature is in excess of 2 billion °K, then the combustion of silicon leads to the formation of the so-called "iron peak" elements. In the process, photons are capable of dislodging alpha particles and protons from the nuclei of heavy elements. These alpha particles and protons then attach themselves to other nuclei, forming heavier nuclei.

Further increases in temperatures up to $8 \cdot 10^9$ °K may result in the formation of almost all the other kinds of heavy chemical elements. Thus, these local "furnaces" in the hydrogen plasma produced different hot clusters of heavy ionized elements. It's possible that some heavy elements had been preserved in the plasma since the stage of hydrogen evolution.

At temperatures above 10 billion °K, complex atomic nuclei become unstable, with disintegration predominating over formation. That is why the primordial plasma basically consisted of hydrogen (80 to 90% by volume), with helium in second place (6 to 10%). The heavy nuclei of all the other elements were present in varying but very small proportions.

In the final analysis, all elements with weight were formed out of weightless physical energy in that same efficient proportion which is necessary for our life. For example, if there were more iron than hydrogen and helium, then we would not be here at all. From this arises the fully relevant question: who created the laws that formed these material elements precisely in the necessary proportions? And why? We call the creator of these providential and efficient laws the Absolute God.

According to Su Hsiu Huang's model published in 1965, these hot and heavy clusters were "shot" from the active regions of the core to the periphery of the plasma cloud at very high speeds. The resistance of the environment would gradually reduce these speeds. As they traveled in a spiral pattern along the magnetic field lines, the heavy clusters would continuously increase their rotational momentum, causing the plasma sphere to rotate more slowly. If the ejected clusters of heavy elements ultimately managed to "break away" from the magnetic field lines, they would carry off a great deal of the total angular momentum with them. Hence the rotational velocity of the condensing cloud of hydrogen plasma would become quite insignificant.

The action of centrifugal force would cause the spherical crust of the plasma sphere to flow into the zone of an equatorial belt, forming a flattened equatorial disk which would extend billions of kilometers from the center of the entire cloud. Rings would separate successively from this plasma disk. The heavier clusters ejected outward into the zone of these rings would attract the lighter clusters and become the cores of future planets. For example, coherent particles of iron along with nickel and cobalt formed the core of the future planet Earth. Increasingly lighter crusts piled onto these cores, and all of the material of the rings condensed around these shells.

The periodic motion of these bodies around the center of the entire plasma cloud was stabilized on a certain orbit at that moment in time when the gravitational forces of attraction were balanced out by centrifugal forces. That is how cosmic bodies consisting not only of hydrogen but of carbon, nitrogen, oxygen, and heavier elements became separated from a fiery ball of hydrogen plasma. These cosmic bodies subsequently became known as Mercury, Venus, Earth, Mars, Jupiter, Saturn, Uranus, Neptune, and Pluto (see table). The center of the fiery ball which remained after these planets had separated from it is referred to as the *proto sun*, because it contained an encoded program of evolution making it inevitable that it would be transformed into the sun. The proto sun primarily consisted of hydrogen plasma and only contained a small amount of the total angular momentum.

If the rate of increase of the rotational velocity of the primordial cloud of hydrogen plasma had not been limited by anything, its circumferential velocity at the equator could have reached the critical level of 299,792 k/sec. But in reality this did not occur. We know, for example, that the equatorial rotational velocity of our sun about its axis is only 2 km/sec. This means that the hydrogen plasma cloud lost much of its angular momentum, largely because of its magnetic field lines.

According to the scientific model of Alfven, the Swedish physicist, these magnetic field lines served as a "transmission belt" that transmitted much of the angular momentum to the planets separating from the proto sun. That is why the primordial cloud was plasma and not gas. If it had been a gas, and not plasma, there would be no magnetic field. If there were no magnetic field lines, the circumferential velocity of our sun at the equator would now be 100 km/sec (and not 2 km/sec).

As Joseph Shklovsky's calculations have shown, 98% of the angular momentum of the entire Solar system was carried off by the planets because of the fact that the primordial cloud consisted of magnetized plasma. Thus, according to Laplace's generally accepted model, all of the planets, and consequently, the earth, "formed before the sun," approximately 5 billion years ago.[3] This was followed by the beginning of the geological evolution of the entire spheroidal surface of the earth, resulting in mountains and depressions. Later on water flowed into these depressions, forming the seas and the oceans.

Thus, while verse 10 in Genesis uses the word "earth" for the dry land and the word "seas" for the gathering together of the waters, in verse 9 the expression "the waters under the heaven" implies the cloud of hydrogen plasma from which our solar system originated. In modern scientific language, the Biblical expression "Let the waters under the heaven be gathered together unto one place," means: "Let this cloud of hydrogen plasma be condensed evenly and form a flattened equatorial disk." And the Biblical expression "and let the dry land appear," means: "Let pieces of heavier (nonhydrogen!) elements separate from the bulk of the hydrogen plasma." In this case the word "dry land" means not just the earth but all the planets, which can no longer be called hydrogen.

It has been scientifically proven that the seas and oceans formed on the earth's surface before the sun did. There are different hypotheses concerning the mechanism by which the seas and oceans formed. Some astronomers have suggested that the moon is a piece of the earth's crust which separated from our planet back in that distant era. A gigantic depression was formed at the separation point where the water flowed. This depression filled with water subsequently became known as the Pacific Ocean.[4]

Other scientists, such as Fred Hoyle, believe that Uranus and Neptune "ejected" chunks of ice from their interiors which in some cases fell to the surface of the primordial earth, forming the seas and oceans.[5] Still another group of scientists believed that the earth's atmosphere and ocean were formed by the "evaporation" of water and other gases from the hot and solid material of the primordial earth. This last model is now generally accepted.[6] According to this model, our earth lost its original atmosphere and began to acquire a new atmosphere approximately 5 billion years ago.

Such chemical elements as hydrogen, oxygen, carbon, and nitrogen had already formed from elementary particles in the primordial plasma. On the primordial earth, hydrogen atoms combined with oxygen atoms to form water ($H_2O$). Carbon atoms combined with oxygen atoms to form the carbon dioxide ($CO_2$) which plants need. By combining with carbon or nitrogen atoms, hydrogen atoms formed methane ($CH_4$) and ammonia ($NH_4$) respectively. Thus, on the primordial earth, hydrogen, carbon, oxygen, and nitrogen combined to form molecules of methane, ammonia, carbon dioxide, and water, which participate in the formation of the simplest forms of biological life.

The earth's primordial atmosphere consisted mostly of hydrogen with much less oxygen. Hence, after water had formed from these elements, the supplies of free oxygen were completely exhausted, and the excess free hydrogen was dispersed into space. Plants absorbed carbon dioxide and gave off oxygen. According to reliable geological and geochemical data, this was why the earth's atmosphere was rich with oxygen 3.5 billion years ago.[7] Consequently, plants appeared on the earth long before this time. As calculations performed by the American scientist Carl Sagan have demonstrated, the simplest forms of biological life emerged on the earth 4.4 billion years ago, i.e. almost immediately after the earth formed, even though the proto sun still hadn't been completely transformed into the sun in that distant era.[8]

Thus, *modern science has confirmed (not discredited!) the Bible's teachings concerning the origin of plants on the earth prior to the final formation of the sun. This hypothesis has been criticized not by natural scientists, but by unscientific atheists. These two groups of individuals are very different, despite atheism's attempts to equate them.*

In essence, biological evolution began back when the first elementary particles of hydrogen plasma, those very same elementary particles in which the ideal program for the inevitable emergence and development of living organisms was encoded, made their appearance. Back in 1907, the renowned Swedish chemist Svante Arrhenius advanced the hypothesis that the "seeds of life" were sown on the earth from other worlds. Now Arrhenius's hypothesis has become a solid scientific theory. *Panspermia* is the science of the origin of biological life from "seeds" in which an ideal program of biological evolution is encoded at the level of elementary particles or even the photon level. These purely energetic (weightless and nonsubstantial) "seeds" could make incredible space voyages, from galaxy to galaxy, from star to star, and from planet to planet. They are not at all afraid of high vacuums, immense pressure, severe cold, high temperatures, the roar of white cosmic holes, the catastrophe of black holes, lethal radiation, or incredible velocities, even if these velocities are equal to or greater than the speed of light. Given the right conditions, they could land on any planet, such as our earth, and grow into living beings.

If we say that human life consists of six successive stages, namely intrauterine formation, birth, development, stability, aging, and death, then we are right on the mark, even though these stages are far from equal in duration and occur at different times for different people. For example, the aging and death of Nicolaus Copernicus (1473-1543) concluded long before the birth of Immanuel Kant (1724-1804), even though death could not occur before birth for the same human being. By perfect analogy, the end of the stage of planetary evolution applies only to our specific solar system and does not in any way imply that planetary evolution has stopped in all other systems. The formation of planets and other (stellar) systems has continued throughout all the subsequent phases of the evolutionary development of the expanding

Universe.

On the basis of the latest advances in the natural sciences, we may summarize *the scientific model of planetary evolution* as follows:

*According to the ideal program of material development of the Universe outlined by God, one of the clouds of hydrogen plasma in our Universe was condensed so that denser "pieces" separated from it and formed the planets. ("And God said, Let the waters under the heaven be gathered together unto one place, and let the dry land appear: and it was so.")*[9]

One of those planets was our spheroidal earth, which had depressions into which water subsequently flowed, forming the seas and the oceans ("And God called the dry land Earth, and the gathering together of the waters God called Seas"). The conditions on the primordial earth were quite favorable for the emergence and evolutionary development of life ("and God saw that it was good").[10]

*And God had devised an ideal program of biological evolution. The elements of this program came from another world to the white holes of our Universe. In the white holes the instructions of this ideal program were converted into material codes at the photon level, and in the clouds of hydrogen plasma they were converted at the level of weighty elementary particles. Thus, the photons generated in the white cosmic holes and the elementary particles of hydrogen plasma contained an encoded program of biological evolution "written" by God. We call the material codes of the ideal program of biological evolution "the seeds of biological life." According to this program, grass, the herb yielding seed, and the fruit tree yielding fruit after its kind would grow on the primordial earth. Photons and elementary plasma particles carried these "seeds of life" with them to the earth, and the program of biological evolution written by God was able to unfold on the earth ("And the earth brought forth grass, and herb yielding seed after its kind, and the tree yielding fruit, whose seed was in itself, after its kind"). That was how the foundations for subsequent biological evolution were laid on the primordial earth ("and God saw that it was good").*[11]

*And the earth formed ("and there was evening") and plants appeared on it ("and there was morning"): thus went the third stage of the creation of the Universe by the absolutely perfect God. We call this period of the creation of the Universe the stage of planetary evolution ("the third day").*

So what can atheism offer to counter this scientific model of planetary evolution except its own whims and illusions? It can only use its completely unsubstantiated initial assumption that "life could not have existed on the earth without the sun."[12] Then we might naturally ask where atheism's scientific proofs are. The answer is simple: "scientific" atheism has no scientific proofs and it could not have them! It only has its convenient initial assumptions which hundreds of millions of people have been compelled to believe blindly, even though it is logically impossible to believe them, because the natural sciences have definitely established that the surface of the primordial earth was never cold enough to make life impossible for all (even the simplest) forms of biological life. The primordial earth itself consisted of very hot material. It cooled down gradually, approaching its current stable temperature state.

If the earth formed before the sun did, then this in no way implies that there was no proto sun at the time which was subsequently transformed into the sun. Even though the proto sun was colder than our present sun, it provided a sufficient amount of heat for the surface of the primordial earth, because its surface was closer to the earth than our current sun. Moreover, the natural sciences have also established that *biological life can emerge and develop not only on the earth, not only on planets, but even on meteorites, which have nothing in common with our sun. Even algae impressions have been discovered on many meteorites.*

Corresponding Member of the Soviet Academy of Sciences Joseph Shklovsky wrote the following on this subject: "We have received reports on many occasions of the discovery of oval inclusions in carbonaceous chondrites which bear an outward resemblance to algae spores. These inclusions luminesced when they were exposed to ultraviolet light. Moreover, they also changed color when they were exposed to special reagents used to detect substances of "biological" origin."[13]

Academician A. I. Oparin wrote that "We have every reason to believe that the earth inherited a large supply of biogenic organic substances from space at its very beginning, which, as they subsequently evolved, served as the material for the emergence of living beings."[14]

Modern science uses the term *planetary chauvinism* to describe the notion that life in the Universe could supposedly emerge and develop only on planets. In all probability, Y. M. Yaroslavsky[15] was this sort of planetary chauvinist. But modern scientists are quite distant from "chauvinistic" ideas.

For example, Cocconi, the Italian physicist, has hypothesized the existence of life even at the nuclear level. Manfred Eigen, the author of the modern theory of molecular evolution, has written the following: "In order for a protein molecule to form by chance, nature would have had to run approximately $10^{130}$ tests and expend enough molecules for $10^{27}$ Universes for this purpose. If a protein were constructed intelligently, i.e. so that the validity of each step could be verified by some sort of selection mechanism, this would require only 2000 attempts. Thus, we arrive at the paradoxical conclusion that the program for the construction of the 'primordial living cell' was encoded somewhere at the elementary particle level."

According to the theory of the Soviet academician M. A. Markov, not just life but even intelligent civilizations could exist in an elementary particle. Academician Markov named elementary particles which contain living or intelligent objects *fridmons* in honor of A. A. Fridman, who first drew our attention to the expansion of the Universe.

Thus, primordial elementary particles already contained an encoded program not just for the formation of the chemical elements but for the appearance of plants and other living beings. This means that the first portions of light, i.e. the primordial photons of the white cosmic holes, a long time before the sun appeared carried the "seeds of life" from another (nonmaterial!) world to our Universe, despite the fact that a photon has no weight and no volume. After getting to the primordial earth, these seeds found appropriate conditions, and life was able to develop on the earth. However, this development was not random. It was lawful ("natural," i.e., established by natural laws) and pro- grammed. Moreover, "natural" in no way implies "spontaneous," as the atheistic literature would have it. "Natural" or "lawful" is the term we use to describe development in accordance with the laws of nature, which were created by God under the auspices of the ideal program.

"The purposeful actions of an individual, to the extent that it may be attributed to the properties of its genetic code, are no more or no less purposeful than the actions of a computer responding to different signals as required by its program."[16]

"And God said, Let the waters under the heaven be gathered together unto one place, and let the dry land appear: and it was so. And God called the dry land Earth, and the gathering together of the waters God called Seas.[17] The Biblical model of the third day (stage) of the creation of the world is fully consistent with the scientific theory of planetary evolution, which states that small "pieces" separated from hot clouds of hydrogen plasma and cooled down before the bulk of the plasma, forming the planets. One of those planets was our earth.

The primordial clouds of hydrogen plasma contained an encoded program which provided that the third stage of the evolution of the Universe would be accompanied by the formation of the earth and the other planets. The significance of the existence of the primordial cloud of hydrogen plasma primarily lies in the fact that planetary evolution would have been impossible without it. The initial state of the third stage of the evolution of the Universe was a cloud of hydrogen plasma, and its inevitable final result was the proto sun and the planets, including our earth. But why were the earth and the proto sun necessary?

According to the Bible,[18] the creation of the world ended on the sixth day with the creation of man, who possesses intelligence. All the preceding stages were preliminary stages, with "preliminary" describing not so much the length of each stage but what and in what sequence had to be done and what was done to achieve the ultimate goal. The interval between the third and the fourth stages of evolutionary development was comparatively short, and therefore, if the Bible had proceeded solely on the basis of the time involved, it could have combined these stages into a single stage. However, the Bible clearly distinguishes between these stages and thus emphasizes that first the earth and then the sun were necessary to achieve the programmed goal.

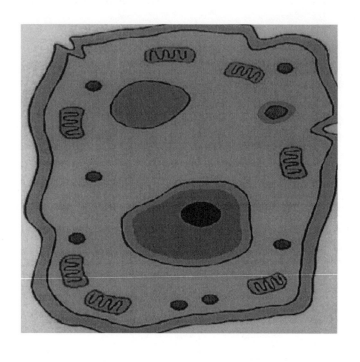

## The "Third Day" (Morning)

According to the ideal program, produced by God and encoded at the elementary level already in clouds of photon and hydrogen plasma, from inorganic matter ("from the dust of the earth") various forms of biological cells were formed on the primordial Earth. In the succeeding stages of evolutionary development, all forms of plants and living organisms were built from these cells. "And God said, "let the Earth bring forth grass, and the herb yielding seed, and the fruit tree yielding fruit after its kind, whose seed is in itself, upon the Earth. And it was so.")

The "Fourth Day" (Evening)

And God created a program of stellar evolution, in accordance with which clouds of hydrogen plasma were transformed into stars and protogalaxies into galaxies. ("And God said, let there be lights in the firmament of the heaven. . . and it was so.")

# =13=

## THE FOURTH STAGE OF CREATION STELLAR EVOLUTION: THE STARS AND THE SUN

**T**he Biblical model in Ancient Hebrew

And God said, Let there be lights in the firmament of the heaven to divide the day from the night; and let them be for signs, and for seasons, and for days, and years. And let them be for lights in the firmament of the heaven to give light upon the earth: and it was so. And God made two great lights, the greater light to rule the day, and the lesser light to rule the night: God made the stars also. And God set them in the firmament of the heaven to give light upon the earth. And to rule over the day and over the night, and to divide the light from the darkness: and God saw that it was good. And there was evening and there was morning, fourth day. [1]

### The Biblical Model in Modern Popular Scientific Language

And God said: Let there be luminaries in heaven and let them be set in relative motion so that winter and summer, day and night may alternate with each other. Celestial bodies must be in periodic motion so that time can be reckoned by them and one era can be distinguished from another, one season can be distinguished from another, and day can be distinguished from night. And let the stars, the sun, and the moon be luminaries in celestial space so as to shine upon the earth, and it was so.

And God made two great luminaries: a larger luminary to rule the day, and a smaller luminary to rule the night and the stars. And God arranged them in space so that their relative motions would make it possible to illuminate and heat the surface of the earth. And so that the seasons of the year (winter and summer) and days and nights would alternate with each other. And God was satisfied with the results of his creation, because within themselves they contained all the conditions necessary for the next stage of the evolutionary development of the Universe, namely the stage of biological evolution. And the stars twinkled ("and there was evening") and the sun shone ("and there was morning"): Thus went the fourth stage in the creation of the Universe by the absolutely perfect God. We call this period of the creation of the Universe the stage of stellar evolution ("fourth day").

## The Modern Scientific Model

In the primordial continuously expanding vacuum space, huge clouds of plasma formed from pure positive energy. This plasma primarily consisted of hydrogen (or antihydrogen) with small admixtures of the other chemical elements. In these clouds, a program was encoded for their subsequent evolutionary development which would inevitably cause them to be transformed into stars and galaxies.

## The mechanism of stars origination

We already know that every such cloud of hydrogen plasma must be simultaneously compressed and rotate around its center, leaving relatively empty space around it.

As compression progresses, the temperature and pressure inside each cloud rise. If the temperature of its core reaches 10 million degrees Kelvin, it will burn hydrogen so as to produce helium and give off a large amount of heat. Thus, the core of the cloud becomes an energy source. The temperature of the plasma decreases from the core to the periphery. That is why the outer layers of the plasma cloud give off their heat into the surrounding space. This kind of plasma cloud is known as a *protostar*. The protostar will continue to undergo compression until equilibrium has been established between the gravitational forces working to compress the protostar and the forces of internal pressure working to decompress it. The protostar will stop undergoing compression and will turn into a star when this equilibrium becomes stable. *This is how stars are born and start shining*. Thus, a *star* is a stable celestial body which gives off light. Stars exist in a stable condition for billions or even tens of billions of years.

## The mechanism of the formation of the galaxies

The Universe was formed from a multitude of clusters, each of which consisted of a vast amount of primordial clouds of hydrogen plasma. These clouds, revolving around one and the same common center, experienced centrifugal forces. At the same time, all of them were attracted to each other and, consequently, to their common center as well, by means of gravitational forces.

In addition, the law of universal gravity also gives all hydrogen plasma clouds the same tendency to stick together and not to be dispersed into the surrounding space. Hence all hydrogen plasma clouds, which cannot break away from their common center, must move in two directions: on a circular orbit and on the radius toward the center.

According to the hypothesis of the Soviet scientist N. S. Kardashev,[2] interstellar (galactic) magnetic lines serve as the mechanism by which a substantial portion of momentum is transferred from each hydrogen plasma cloud to the entire system of clouds which forms one common galaxy. As a result of this, the vast galaxy is imparted rotation around its galactic center at a limited velocity, and the rate of increase in the velocity of rotation of each individual hydrogen plasma cloud is substantially slowed.

A system of such protostars or hydrogen plasma clouds is known as a *protogalaxy*. A protogalaxy will continue to undergo compression until equilibrium is established between the gravitational forces working to "compress" the protogalaxy and the centrifugal forces working to "decompress" it. A protogalaxy will stop undergoing compression and will turn into a galaxy as soon as the equilibrium between these forces stabilizes. Thus, a *galaxy* is a cluster of a large number of stars (or protostars) which form a single cosmic system. The set of every single galaxy constitutes our Universe, which has a single coherent three-dimensional space. The entire Universe contains a large (but not infinite!) number of galaxies.

Every galaxy rotates about its own galactic center and simultaneously travels on a spiral orbit about the center of the Universe at different velocities. If the Universe did not experience expansion and compression, then the orbital motion of a galaxy about the center of the Universe would be circular, not spiral. The time interval in which a galaxy completes one full rotation about its own axis is known as a *galactic day*. Each galaxy has its own day which differs from the days of the other galaxies. The time interval in which a galaxy completes one full revolution about the center of the Universe could be called a *galactic year*. Then each galaxy would have its own galactic year which would differ from the years of the other galaxies. But according to Joseph Davydov's theory, the Universe will only rotate 180 degrees about its axis during its entire existence: during the first fourth of the rotation the Universe was born and expanded, while during the second fourth of the rotation it will be compressed and disappear. Thus, the period of existence of the Universe is only equal to half of its cosmic day, and in the second half, an Anti-Universe will be born, expand, undergo compression, and disappear instead of the Universe.[3] In contrast to all the other galaxies, the name of our Galaxy is capitalized. Our Galaxy has a spiral structure and consists of 150 billion stars. Its diameter is approximately $10^{18}$ kilometers, i.e., it would take a ray of light 100,000 terrestrial years to travel from one end of our Galaxy to the other.

All stars rotate about their own axes and simultaneously travel on their own orbits about their galactic center at different circumferential velocities. The time interval during which a star completes one full rotation about its own axis is known as a *stellar day*. Each star has its own day which differs from the days of the other stars. The time interval in which a star completes one full revolution about the center of its own galaxy is known as a *stellar year*. Each star has its own year, which differs from the years of the other stars.

Our sun, which originated from the protosun, is one of the large number of stars in the Universe. The protosun continued to undergo compression even after the planets separated from it. As it underwent compression, according to the laws of mechanics, the protosun had to start rotating faster and faster, but the magnetic fields of the other stars kept it from doing so. Interstellar magnetic lines imparted rotational momentum from the protosun to the other stars and thus to the entire Galaxy like the spokes which transmit motion from the axle of a wheel to its rim. This compression continued until equilibrium was established between the gravitational forces working to compress the protosun and the forces of internal pressure working to decompress it. The protosun stopped undergoing compression and turned into the sun once this equilibrium became stable. The sun and the nine planets (Mercury, Venus, Earth, Mars, Jupiter, Saturn, Uranus, Neptune, and Pluto), which travel on their own orbits about the sun, form the *solar system*, whose radius is approximately 6 billion kilometers. Thus the *sun* is the star about which our earth travels on an orbit. The sun is 1.39 million kilometers in diameter.

The sun primarily consists of hot hydrogen plasma. The temperature on its surface is roughly 6000 degrees Kelvin. However, temperatures in the central layers of the sun are much higher than the surface temperatures, reaching levels of 20 million degrees Kelvin. Such high temperatures are accompanied by thermonuclear reactions by which hydrogen protons are transformed into helium nuclei. The energy liberated in the process slowly seeps from the inner to the outer layers of the sun and is then irradiated into the surrounding space. The heat transfer, shape, and dimensions of the sun have not changed over the last 4.5 billion years. Scientists have hypothesized that the sun's energy will be exhausted in 10 billion years. As a result, in the distant future the sun will start to cool down gradually and will ultimately turn into a cold dark body.

The motion of the sun, like the motions of all other stars, is a combination of translational and rotational motions. The sun rotates about its own axis and simultaneously travels forward on a galactic orbit at a velocity of 250 kilometers per second. The time interval during which the sun completes one full rotation about its own axis is called a *solar day*. The time interval during which the Solar System completes one full revolution about the center of the Galaxy is called a *solar year*. One solar year contains 200 million terrestrial years.

All nine planets of the Solar System have a slightly oblate spherical shape. They are in a state of continuous rotation about their own axes, and continuously travel in orbit around the sun. The orbits on which the planets travel around the sun are elliptical in shape. However, these ellipses differ from circles very slightly. The orbits of the planets lie very close to a common plane, which almost coincides with the plane of the sun's equator. This orbital motion of every single planet occurs in the same direction, which coincides with the direction of the sun's own rotation about its axis. All the planets, except Venus, rotate in the same direction about their own axes.

Venus' rotation about its own axis is not only opposite to that of the other planets of the Solar system. It is also opposite to the direction of Venus' own revolution about the sun. In contrast to the hot sun, its planets are relatively cold bodies. The basic characteristics and dimensions of the Solar System are given in the table on the next page.

The earth on which we live is one of the nine planets of the Solar System. Its mass is approximately $6 \cdot 10^{24}$ kg. The average radius of the earth is 6,370 kilometers. The alternation of day and night on the earth is the result of our spherical planet's rotation about its own imaginary axis passing through two points on the earth's surface known as terrestrial poles. If it is day on the half of the earth illuminated by the sun, it is night on the other (unilluminated) half. The time interval in which the earth completes one full rotation about its own axis is called a *terrestrial day*. One terrestrial day includes 24 hours of terrestrial time. According to the special theory of relativity, time depends on a body's velocity. Hence terrestrial time should never be confused with the time which takes place on other celestial bodies traveling at different

## The Planets of the Solar System

| Planet | Average distance from Sun, km | Diameter of planet, km | Mass, kg | Average density, g/cm³ | Period of rotation around its own axis | Orbital period | Orbital velocity km/sec | Average temperature of illuminated surface, °C |
|--------|------|------|------|------|------|------|------|------|
| Mercury | 58 | 4,840 | $3\,3 \cdot 10^{23}$ | 3.8 | 58 days | 88 days | 48 | 340 |
| Venus | 108 | 12,228 | $4.9 \cdot 10^{24}$ | 4.8 | 243 days | 225 days | 32 | 380 |
| Earth | 150 | 12,740 | $6\,0 \cdot 10^{24}$ | 5.5 | 24 hours | 365 days | 30 | 20 |
| Mars | 228 | 6,770 | $6.4 \cdot 10^{23}$ | 3.9 | 24 hours | 687 days | 24 | -10 |
| Jupiter | 778 | 140,720 | $1.9 \cdot 10^{27}$ | 1 3 | 10 hours | 11.8 years | 13 | -130 |
| Saturn | 1428 | 116,820 | $5.7 \cdot 10^{26}$ | 0.7 | 10 hours | 29.6 years | 9 6 | -150 |
| Uranus | 2872 | 47,100 | $8.7 \cdot 10^{25}$ | 1 5 | 10 hours | 88.7 years | 6.5 | |
| Neptune | 4498 | 44,600 | $1\,0 \cdot 10^{26}$ | 2.5 | 12 hours | 166 years | 5.4 | -160 |
| Pluto | 5,910 | 6,000 | $5.0 \cdot 10^{24}$ | 4.8 | | 248 years | 4.8 | |

velocitities.

As it rotates about its own axis, the earth also travels around the sun at a velocity of 30 kilometers per second on an orbit whose radius varies very little, from 147 to 152 million kilometers. The time interval during which the earth completes one full revolution around the sun is called a *terrestrial year*. One terrestrial year is equal to 365.26 terrestrial days on average.

Each planet, each star, and each sun has its own days and years. And under no circumstances can they be confused. For example, one solar day is equal to approximately 26 terrestrial days. This means that it takes the sun approximately 26 terrestrial days to complete one full rotation about its own axis. In the process we should keep in mind that different belts of the gaseous sun rotate about its axis at several different velocities. However, this difference is not always significant.

Neither terrestrial time nor terrestrial days should ever be confused with the "days of the creation of the World" mentioned in the Bible. We know that at the time of the birth of the Universe all of its elements were moving relative to one another at velocities which were equal (or almost equal) to the speed of light. We also know that our planet earth is now traveling on its circular orbit at a velocity of only 30 kilometers per second. According to the special theory of relativity, time depends on velocity. Hence preliminary estimates would indicate that *one Biblical day would be roughly equal to 2 billion terrestrial years*. Atheism has intentionally confused these concepts in order to divert the masses from the sacred truth and arouse them against the Bible.

If the earth did not rotate about its own axis, we would not be able to reckon days or distinguish day from night or light from darkness. If the earth did not revolve around the sun, then we would not be able to reckon years. If the moon did not revolve around the earth, we would not be able to reckon months.

The motion and configuration of the planets and their satellites in the solar system is the most precise, expedient, and useful for our existence and development. The precision of their relative motion guarantees the stability of our system for tens of billions of years. The human mind cannot fathom such precision. No artificial satellite could be made so reliable. The comprehensive expedience of all the elements of the Solar System has dazzled the minds of scientists and has required them not only to admit but to praise the lofty intelligence of their brilliant Creator.

For example, the earth's own axis of rotation forms an angle of 66.5 degrees, not 90 degrees, with the plane of its orbital motion around the sun. And this is no coincidence, because this incline of the earth's terrestrial axis allows the earth to turn its northern hemisphere and then its southern hemisphere toward the sun. In the first case it will be the summer in the northern hemisphere and the winter in the southern hemisphere. In the second case it's the opposite, i.e., the winter in the northern hemisphere and the summer in the southern hemisphere. If the earth's own axis of rotation were to form a right angle with the plane of its revolution around the sun, then no change in seasons would take place on the earth. It would be equally cold year round near the poles, and it would be equally hot near the equator. Then life would become unbearably hot in some parts of the world and unbearably cold in other parts. At best this circumstance would significantly reduce the surface of the earth suitable for life. More probably, our planet would become unsuitable not only for the development of living beings but for their genesis. Hence the incline of the earth's own axis of rotation to its plane of orbital motion around the sun (like all the other optimal characteristics of the Solar System) is supremely expedient and is profoundly meaningful. This kind of expedience and foresight could not be pure happenstance and is most likely the product of great intelligent creativity.

Another example of programmed expedience in the Solar System is the purpose of the moon, which is the earth's natural satellite and revolves around the earth at a velocity of 1 kilometer per second on its orbit. The radius of this orbit is approximately 384,000 kilometers, while the radius of the moon itself is 1738 kilometers. This means that the dimensions of the sun (the diameter of the "greater light") are 400 times greater than the dimensions of the moon (the diameter of the "lesser light"). Thus, the author of the Bible, who lived in the age of bows and arrows, knew 3300 years ago that the sun was much larger than the Moon in size, even though they seem approximately the same size to the naked human eye on earth. Scientists became aware of this relatively recently, even though modern technology can launch rockets into space. The only thing we can do is praise the ancient arrow which overtook the modern missile. The moon's period of revolution around the earth is equal to 28 terrestrial days.

The moon's mass is only $73 \cdot 10^{21}$ kg. This relatively small body could not remain hot for very long and serve as a light source. Hence the surface of the moon is dark. But the moon's motion is of a complexity which allows it to illuminate the earth at night by reflecting sun rays. Even though the moon illuminates the earth by reflecting sun rays instead of its own light, this does not mean that the moon cannot be called a "light," no matter how the atheists would like to portray it.

If there were no stars or moon, then night on the earth would be so dark and frightful that humans and many living creatures would have found navigation impossible (certainly until humans learned to make fire). It would be doubtful that people could have survived at all under these conditions. Hence the existence of the moon on a near-earth orbit has a special meaning, because at night the moon reflects sunlight onto the earth and thus affords people and animals relative freedom of navigation. Consequently, the genesis of the moon was expedient and programmed, not accidental, even though there is no scientific unanimity with respect to its physical origins. Some scientists believe that the moon is a broken-off fragment of the earth. Other scientists believe that the earth "took the moon prisoner" as the moon was passing by. There are also many other hypotheses.

Astronomers who have tried to unlock the secrets of the structure of the Solar System have become convinced that the Solar System did not develop solely on the basis of purely mechanical or physical laws. It prepared all of the optimal conditions necessary for the birth and development of life on the earth in advance. Any optimization can only be the result of a program. Any program can only be the product of intelligent creativity.

Under the onslaught of scientific facts, atheism has been compelled to admit that the stars, the sun, the moon, and the other planets originated as a result of logical, programmed, purposeful, expedient, and evolutionary development, and not by accident, and consequently not by themselves. This program provided for everything ahead of time.

For example, if the earth were somewhat more distant from the sun, like Mars, or somewhat closer, like Venus, it would not be possible for you and me to exist. The sun sends exactly as much solar and thermal energy to the earth as is needed for biological evolution, no more and no less.

The Soviet scientist Ephraim Levitan writes: "The harmony and order in the motion and configuration of the planets and their satellites is a source of great aesthetic pleasure and has dazzled everyone who is now trying to unlock the secrets of the structure of the Solar System."[4] Allow me, then, to ask: exactly who established this order and who guaranteed this harmony?

To think that harmony and order reigned by accident is to believe in atheist miracles and fairy tales. To think that all the dimensions, configurations, and motions of the planets were established by accident exactly the way we need them without a predetermined purpose is to descend from the heights of science into the stagnant swamp of atheist superstition. The most advanced cybernetic system in the Universe is the Universe itself. Hence we would have to call God, who created the Universe, the best intelligent designer in the world.

We know that a tropical year on the earth is equal to 31,556,925.9747 seconds. With each passing year it only gets $10^{-4}$ seconds longer.[5] This precision of the earth's travel in orbit will guarantee its stability for the next 10 billion years. If modern scientists were billions of times more intelligent than they are now, even then they would not be able to devise a mechanism so precise. But this sort of precision exists! It exists outside of our mind and independently of us! It not only exists on its own, it also guarantees the possibility of our existence.

To think that this sort of precision and expedience in the organization of the planets arose all by itself without any sort of an outside intelligent creator is much worse than believing that television came into being all by itself without any engineers. But atheism not only believes in this fairy tale itself, it has propagated this atheistic superstition throughout the world, speaking in the name of science. Thus, atheism, which formally rejects fairy tales, in reality believes in them. It is not religion but atheism which relies on blind faith in such fabricated fairy tales as natural laws without a legislator, a program of development without a programmer, a product of intelligent creativity without a creator, and so forth.

One is correct in saying that human life consists of six successive stages, namely intrauterine formation, birth, development, stability, aging, and death, even though these stages are in no way equal in duration and even though they occur at different times for different people. By perfect analogy, the end of the stage of stellar evolution only applies to a specific part of our Universe and in no way means that stellar evolution has ceased altogether in all the other parts of the Universe. The formation of planets, stars, and galaxies has continued in other parts of the expanding Universe and in all subsequent stages of its evolutionary development.

We can observe the birth of new galaxies in white holes even now, using quasars as an example. *Quasars* are small superpowerful energetic cores located on the periphery of our Universe and receding from us at velocities close to the speed of light. The radius of this core is approximately one-fifth to one-sixth that of the Solar System, even though it emits millions of times more energy than our sun, creating the impression of a "colossal explosion." The source of the colossal positive energy found in quasars is the white holes, in which equal amounts of positive and negative energy are generated from nothing in full harmony with the law of the conservation of energy. The negative energy generated in the white holes of quasars is transformed into the invisible seas of energy of the spatial vacuum, which are expanding the boundaries of physical space. In the future, today's quasars will be transformed into entire galaxies or even clusters of galaxies.

We currently see those quasars that constituted the Universe's boundary back when the Universe was 6-7 billion earth years old. It has been taking their light another 6 billion years to travel to us. They ceased to exist long ago, and many of them have evolved into galaxies and stars. The present boundary of the Universe is 12-14 billion earth years away from us. Therefore "we can see it" only in 12 billion years, if we still live on the planet at the time.

The natural sciences have firmly established that any galaxy recedes from an observer at a velocity directly proportional to the distance between them. The farther the galaxy is from the observer, the faster it is receding. The most remote points of the Universe, which can be viewed by means of telescopes, such as quasars, are receding from terrestrial observers at a velocity equal to 90% of the speed of light.

It is said that the expansion of the Universe occurred in two stages, an explosive stage and a gradual stage. This definition is not quite accurate, because the periphery of the Universe is always receding from the center at the maximum possible velocity. But because the periphery was at the center in the primordial Universe, the entire Universe expanded at the speed of light, creating the effect of a colossal explosion ("the big bang"). Now there are many galaxies in the Universe with heavy cores. Hence, expansion at the center of the Universe cannot proceed as rapidly as it does on the periphery.

Over the 70 years of existence of the Soviet Communist Party, all of totalitarian atheism's attempts to rescue itself from a complete scientific disaster ended in failure. The Communist superpower couldn't have built up its military might without the development of natural sciences such as physics and astronomy. The development of the natural sciences was inevitably accompanied by the triumph of truth over falsehood, meaning that atheism lost its foothold little by little.

First it was proven that the Universe is expanding and is in no way infinite. Then the atheistic fairy tales of the infinite density of the zero volume of the primordial Universe collapsed. Subsequently atheism was forced to admit that energy could arise from nothing. Ultimately "dialectical" materialism exhausted itself, and "scientific" atheism, which was the ideological soul of the Soviet Communist Party, croaked completely. We know that a body cannot live without a soul. Hence the largest and mightiest power in the world, once it lost its spiritual ideology, collapsed like a house of cards, exactly at the time when its military might have reached its peak.

It was not atheism that collapsed as a result of the disintegration of the Communist superpower, it was the Communist superpower which collapsed because of the utter scientific bankruptcy of totalitarian atheism. The "ideas of Communism" were the first to emerge, and only afterwards was the Communist superpower created. The decline of Communism followed the same basic pattern: the scientific collapse of the "ideas of atheism and Communism" came first, followed by the destruction of the physical power of the Communist Party of the Soviet Union. Thus, in spite of its unfounded assertions, the Communist Party of the Soviet Union provided a demonstration of the *primacy of the idea and the subordinacy of matter* for the entire world.

The Communist Party of the Soviet Union, which was a primary source of atheistic misinformation, is now defunct. But atheistic superstition continues to exist by inertia. The source of the infection has disappeared, but the disease itself still exists not just in Russia but far beyond its borders. Certain scientists who were brought up on totalitarian atheism are still looking for ways to rescue atheism from total scientific disaster. One such attempt was the hypothesis of "natural selection in the world of Universes."[6] According to this hypothesis, black holes are converted into white holes, from which Universes are selected and formed independent of any Creator.

The hypothesis of "natural selection in the world of Universes" is pure science fiction and in no way scientific, because the transformation of black holes into white holes seems just as impossible as the transformation of a dead man's corpse into the live body of a baby. If the mass of a Universe of supercritical density were two or more times greater than the mass of the sun, no kind of natural selection would be able to save this Universe from total collapse and disappearance. That is why the genesis and expansion of Universes is possible only from ideal zero, and is not possible from small material "cosmic eggs." If the mass of a Universe were equal to or less than twice the mass of our sun, then energy would have to be created out of nothing, which cannot occur spontaneously, in order to transform this Universe into a Universe like ours. But the white holes on the periphery of our Universe are the sources of new galaxies, *not new Universes*, as the atheists have suggested. These white holes are part of our Universe and do not exist outside of it.

A scientific analysis of the structure of the Universe indicates that all its elements and systems are similar models of the "planetary" type. This fact also provides evidence that the Universe was created according to the consistent expedient design of one and the same highly intelligent Creator.

If the structure of white holes on the periphery of the Universe is reminiscent of the structure of a living cell to the same extent that the structure of the solar system resembles that of an atom, this only indicates that the entire world has developed on the basis of a consistent program and the coherent design of one and the same Creator and does not mean that Universes are "selected" naturally, by themselves, without any Creator.

Even though natural selection does exist, it in no way implies that living beings or Universes are born and die spontaneously, without any

Creator. Any natural selection conforms to certain laws, and the complete collection of all the laws of nature constitutes a coherent program for the comprehensive development of the Material World. But laws do not exist without legislators, and an expedient program cannot exist without an intelligent Creator. To think that natural selection occurs without any Creator is tantamount to believing in the spontaneous origin of an automatic control system without any creative thought or any engineer.

On the basis of the latest information from the natural sciences, we can succinctly formulate a *scientific model of stellar evolution* as follows:

*And God created an ideal program for the material development of the Universe, by which clouds of hydrogen plasma would inevitably and naturally be transformed into the stars and the sun, the planets and the earth, and the satellites and the moon. All of these celestial bodies would have to be in periodic motions by which one could reckon time and distinguish one era from another, one season from another, and day from night. "And God said, Let there be lights in the firmament of the heaven to divide the day from the night, and let them be for signs, and for seasons, and for days, and years."[7]*

*According to this program drawn up by God, the sun would send exactly as much light and heat energy to the earth as would be necessary for biological evolution, and no more or no less. The stars in the heavens would have to be bright enough for their light to reach the earth. And subsequently everything happened exactly as it had been programmed previously. "And let them be for lights in the firmament of the heaven to give light upon the earth, and it was so."[8]*

*According to the program drawn up by God, the clouds of hydrogen plasma were transformed into stars, the stars combined to form galaxies, and the galaxies made up the Universe. One such star is our sun ("the greater light"), which illuminates the earth only in the daytime. The moon ("the lesser light") also originated in our solar system and together with the stars illuminates the earth only at night. "And God made two great luminaries: a larger luminary to rule the day, and a smaller luminary to rule the night; God made the stars also."*[9]

The motion of every celestial body is a combination of translational and rotational motions. The moon travels on a circular orbit around the earth. The earth rotates about its axis and travels forward on an orbit around the sun. The moon executes a complex form of motion which allows it to illuminate the earth at night by reflecting sun rays. The sun rotates about its own axis and simultaneously travels forward in an orbit around the center of our Galaxy. Each star rotates about its own axis and simultaneously travels forward on an orbit around the center of its own galaxy. Each galaxy rotates about its own axis. According to the theory of relativity, any motion of the planets is relative, and not absolute. This means that any celestial body (the earth or the sun) may be used as a fixed system of reference. If we say that the earth revolves around the sun, we could likewise say that the sun revolves around the earth. In this lies the essence of the generally accepted theory of relativity.[10]

*"And God set them in the firmament of the heaven to give light upon the earth."*[11]

The time interval in which the earth completes one full rotation about its own axis is called the terrestrial day. One terrestrial day contains 24 hours of terrestrial time. The time interval in which the earth completes one full revolution around the sun is called the terrestrial year. One terrestrial year on average equals 365.26 terrestrial days. The time interval in which the sun completes one full rotation about its own axis is called the solar day. The time in which the sun completes one full revolution around the center of the Galaxy is called the solar year. There is no watch in the world which runs more accurately than the planetary and solar systems. Summer and winter, day and night, alternate and change places.

*"And to rule over the day and over the night, and to divide the light from the darkness."* And God was satisfied that the relative configuration and motion of the earth, sun, and stars contained all the conditions necessary for biological evolution. *"And God saw that it was good."*[12]

*And the stars twinkled ("and there was evening") and the sun shone ("and there was morning"): thus went the fourth stage of the creation of the Universe by the absolutely perfect God. This period in the creation of the Universe is called the stage of stellar evolution: ("fourth day").*

Above we gave a purely scientific model of the fourth stage of the evolutionary development of the Universe. So how does it differ from the Biblical model? By analyzing and comparing them, we become convinced that there are no essential semantic differences between the two. The only difference lies in the form and style of presentation and in the distinctive features of the ancient and modern languages. The scientific model corroborates the Biblical model instead of discrediting it.

So what can atheism use to counter this scientific model of stellar evolution besides its own whims and illusions? It can only counter the scientific model with its own anti-scientific and ridiculous fabrication that the world was supposedly created by a common worker, and not God.[13] But then we might appropriately ask where the scientific proofs of atheism are. And the answer is quite simple: "scientific" atheism has no scientific proofs nor could it!

Modern science corroborates the Bible instead of discrediting it. Science has rejected atheism, not religion. That is why atheism is making desperate attempts to turn the helplessness of its ideology against the Bible as well as science. *Atheism has not only slandered but has also unforgivably distorted the contents of the Bible in the eyes of the masses. Atheism has fabricated lies, attributed these lies to the Bible, criticized its own lies, and has used this criticism as the basis for stating that science supposedly discredits the Bible.* The lavish and completely unlimited funding for atheist projects, unabashed distortion of the contents of the Bible and the scientific facts, and shameless slanders have more than compensated for the lack of scientific facts in atheism's beliefs.

According to the Bible, God is an absolutely perfect ideal category with no material attributes. "Material gods" and idols are characteristic of atheism (and not the Bible), because only atheism blindly believes in the creative potential of dead matter. To believe that dead nature created the supremely efficient structure of the Solar System is tantamount to deifying unintelligent nature. The atheistic superstition of the creative powers of unintelligent nature is essentially a form of antiscientific idolatry where all unintelligent matter as a whole, instead of some statue, serves as the deified idol.

In contrast to atheism, the Bible considers a nonmaterial God with an absolutely free will and an absolutely perfect intelligence to be the creator of the expedient structure of the Universe. The God Elohim created the laws of nature and devised an ideal program by which the Universe would arise from nothing and develop, forming the stars and the galaxies, the planets and the sun, the earth and the Moon. God created the Universe by means of the ideal program encoded in the elementary particles of the white holes and hydrogen plasma, not by such physical means as a hammer and nails, as described in one Soviet distortion of the story of creation, in which the Hebrew God is an old man hammering the stars into a dome-like sky. Such a primitive notion of the creation of the Universe and of God is characteristic of atheism, not the Bible. To think that an ideal God with no material attributes created the Universe by nailing lights to a wooden dome with a hammer is equivalent to confusing the mind of an engineer with a particular component of a program-controlled computer system.

The Ideal God created the moon, the earth, the sun, the stars, the galaxies, and the entire Universe not by means of a hammer and nails, but by means of the "Word of God," i.e., by means of an ideal program which was at first encoded at the energetic and elementary levels and then at the atomic and molecular levels. Any law of nature is a specific rule of the comprehensive ideal program created by God and encoded at the energetic level. Any form of matter, starting with an unintelligent quark and ending with the human brain, is absolutely obedient to these laws. Any command from the Ideal World to the Material World, which unambiguously defines the standards of behavior for all material elements and systems, is transmitted by means of weightless elementary particles in which intelligent microcivilizations process ideal information into material codes and vice versa.

"And God said, Let there be lights in the firmament of the heaven to divide the day from the night. . . And God made two great lights; the greater light to rule the day, and the lesser light to rule the night: God made the stars also."[14]

The Biblical model of the fourth day of creation of the world is completely consistent with the scientific theory of stellar evolution, according to which the stars were formed from hot clouds of hydrogen plasma. Our sun is one such star.

A program was encoded in the primordial clouds of hydrogen plasma by which the third and fourth stages of the evolution of the Universe would be accompanied by the formation of the planets and the stars, the earth and the sun. Joseph Shklovsky has written the following on the subject: "The incredible variety of stars, which include neutron stars, planets, comets, and living matter with its incredible complexity and many other features which we still cannot grasp, in the final analysis all developed from this primitive cloud of plasma. We automatically arrive at the analogy of some kind of gigantic gene in which the entire future and unbelievably complex history of matter in the Universe was encoded. . ."[15]

Compare this scientific statement with the following phrase from the Bible: "And God said, Let there be lights in the firmament of the heaven." The significance of the existence of the primordial cloud of hydrogen plasma primarily lies in the fact that stellar evolution in general and the evolution of our Solar System in particular would be impossible without it. The initial state of the third and fourth stages of the evolution of the Universe consisted of clouds of hydrogen plasma and their inevitable end result was the galaxies, the stars, and the planets, including the sun and the earth. But for what purpose were the earth and the sun necessary?

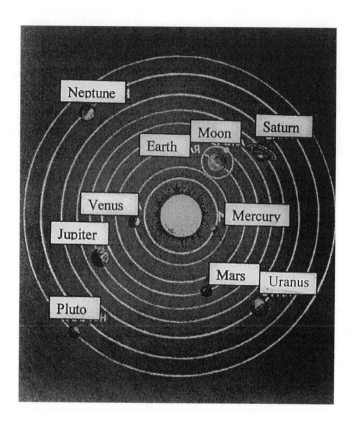

The "Fourth Day" (Morning)

According to the program stellar evolution, produced by God, one of the clouds of hydrogen plasma was transformed into our Solar System. ("And God made two great lights, the greater light to rule the day, and the lesser light to rule the night. God also made the stars.")

The "Fifth Day" (Evening)

And God created an ideal program of biological evolution, in accordance with which from the simplest forms of biological cells "the earth brought forth grass, and herb yielding seed after its kind, and the tree yielding fruit, whose seed was in itself, after its kind."

# =14=

## THE FIFTH STAGE OF CREATION
## BIOLOGICAL EVOLUTION: FISH AND BIRDS

**T**he Biblical Model in Ancient Hebrew
And God said, Let the waters bring forth abundantly the moving creature that hath life, and fowl that may fly above the earth in the open firmament of heaven. And God created great whales, and every living creature that moves, which the waters brought forth abundantly, after their kind, and every winged fowl after its kind: and God saw that it was good. And God blessed them, saying, Be fruitful, and multiply, and fill the waters in the seas, and let fowl multiply in the earth. And there was evening and there was morning, fifth day.[1]

### The Biblical Model in Modern Popular Scientific Language
And God said: let the fish first appear in the water and then the birds in the skies above the earth. And God first created living creatures in the water, then the large fish and marine animals, and then God created the winged birds which fly through the air. All of them were born, developed, matured, produced offspring after their kind, aged, died, and were born again. And God was satisfied with the results of this creation, because the biological systems contained all the conditions necessary for the next stage of the evolutionary development of the Universe, namely the stage of intelligent evolution. And God blessed them, saying: be fruitful and multiply and fill the water in the seas and the space above the earth. And thus fish arose ("and there was evening") and the birds arose ("and there was morning"): thus went the fifth stage of the creation of the Universe by the absolutely perfect God.

This period in the evolution of the Universe is what we call the stage of biological evolution: ("fifth day").

## The Modern Scientific Model

And God created an ideal program for the material development of the Universe by which hydrogen plasma clouds would naturally be transformed into the stars and sun, the planets and the earth, and the satellites and the moon. In the process it would be quite appropriate to ask how life appeared on the earth. Many scientists believe that "this extremely important problem of the modern natural sciences has not yet been resolved."[2] Nevertheless, modern science still holds to an evolutionary model of the origins and development of life. But what does this mean? Let us try to answer this question in a concise way.

*Biological evolution* is what we call the logical and programmed process of the gradual transformation of inanimate matter into living organisms, which is accompanied by radical and qualitative changes which occur in stages. If we call evolution a "natural" process, this in no way means that biological evolution occurs spontaneously, by itself, without any program, without any goal, without any will, or without any Creator.

On the contrary, we call biological evolution a "natural" process because it is fully consistent with the laws of nature, the comprehensive collection of which constitutes the ideal program of biological evolution. This program clearly dictates the necessary transformation of inanimate matter into living organisms in this exact way and in this way only. In the process we should not forget that laws cannot exist without a lawgiver, and that a program cannot exist without an intelligent Creator.

The ideal program for the material development of our Universe was created by the absolutely perfect God in another (ideal, not material!) world before the birth of the Universe itself. The completion of the program was followed by a qualitative transformation of the nonexistence of the Universe into its existence in the following way: initially there was the first white hole (a zero point) at which our Universe originated and began to expand in the form of a zero sum of positive and negative energy. The rules of the ideal program came to our material from another (nonmaterial) world through the white holes, where ideal information was converted into energetic codes at the level of weightless, bodiless, and shapeless photons.

The elements of these energetic codes, which at the photon level corresponded to the ideal program of biological evolution, could already be called the *"seeds of biological life."* The period of the incarnation of an otherworldly idea in the form of these "seeds of life" is called the first stage of biological evolution.

These "primordial seeds of life" contained inside the photons only provided the beginning for biological evolution, even though every weightless and bodiless photon, whose physical volume is equal to ideal zero, contains a definite command indicating what the same physical energy must be transformed into, depending on the seed, i.e. tomatoes, cucumbers, plums, and so forth. These codes contained "inside" the photons must be clearly distinguished from the wave codes of physical energy (positive or negative), which have always and continuously carried information from the other (ideal) world to our (material) world, like radio waves.

Subsequently, as positive energy was transformed into clouds of hydrogen plasma, another qualitative change occurred. The energetic codes of the ideal plasma were transformed into substantial codes at the nuclear or electron levels. The period in which purely energetic (weightless and bodiless) "seeds of life" were transformed into substantial (weighty and elementary substantial particle) codes at the nuclear or electron level is known as the second stage of biological evolution.

As we have already mentioned, the ammonia, water, and methane molecules which participate in the formation of life had already arisen on the primordial earth. Under the proper conditions and under the action of the aforementioned seeds of life which had reached the earth, these molecules gave rise to the proteins and nucleic acids which served as the next (third) encoded foundation of material life. In this case, the elementary particle codes which carried information from the other (ideal!) world to our material earth underwent a new qualitative change and were transformed at the molecular level into the codes of proteins and nucleic acids, namely DNA (desoxyribonucleic acid) and RNA (ribonucleic acid).

Protein and nucleic acid (DNA and RNA) molecules then formed biological ("living") cells and thus underwent the next qualitative change. The biological cell was the fourth form of matter which contained the material codes of the ideal program of biological evolution. These biological cells are the "building blocks" of all living beings and all plants and animals, ranging from the tiniest microbe to the largest mammal. Hence we call the expedient coherence of some set of biological cells which gives rise to a new form of matter which contains the material codes of a genetic program a *biological system.*

A *genetic program* is the name we give to an ideal program which will unambiguously determine the behavior of a given biological system or element under any conditions. *Genetic code* is what we call the material copy of an ideal genetic program. The period of the formation of plants from biological cells is called the fifth stage of biological evolution, while the period of the formation of animals from biological cells is called the sixth stage of biological evolution.

Biological systems include all living organisms, ranging from the simplest bacteria to the largest animals, which have appeared on the earth in different eras. According to extant scientific data, bacteria and very simple algae appeared 4.6 billion years ago, very simple marine invertebrates such as the sponge and the jellyfish appeared 500 million years ago, the first birds appeared 200 million years ago, hoofed and predatory animals appeared 70 million years ago, while man appeared only 60,000 years ago (see any biology textbook). This scientific data corroborates rather than discredits the Bible, which states that first plants appeared, followed by birds and fish, and only afterwards by land animals and man. These times apply to our planet. In other solar systems, however, biological evolution may have occurred at a different time, even though in any case it would have to have occurred as the fifth stage following the formation of a suitable planet and suitable sun.

A protein molecule consists of twenty amino acids, which may form an incredibly large number of combinations. The sequence of amino acids in a protein molecule determines the function or normal behavior of the protein. Just as the legislatures of different countries convey the semantic (ideal) content of their laws to all their citizens by means of such material codes as the letters of the Russian, English, or Chinese alphabets, the otherworldly God conveys the semantic content of the ideal program of biological evolution and the laws of nature which determine the behavior of any living organism to all biological systems by means of such biological codes as protein amino acids.

Thus, protein amino acids are the building blocks of a variety of material codes which carry a definite command from God to biological elements and systems for any situation. And God understands and plans these commands, while a biological system carries them out unconsciously, i.e., just like any computer system blindly carries out the conscious will of an engineer or programmer. In this case the protein plays the role of a biological alphabet which consists of a mere 20 amino acids (protein letters).

But a protein cannot exist forever. Like any other material system, a protein is born, develops, functions, ages, dies, and is reborn. As it dies and is reborn, the protein must preserve the code which carries otherworldly information. This purpose is served by the DNA molecule, which has the remarkable property of producing an exact copy of both itself and the living cell which contains it. Thanks to the DNA molecule, any old cell can divide into two new cells, each of which contains a complete and exact code of genetic information. If a DNA molecule were unable to accomplish this, then dying cells could not be reconstructed and biological evolution would prove altogether impossible.

A DNA molecule is a unique four-letter alphabet of biological life, which passes on a genetic code as a legacy from "parents" to "children" and basically consists of the following four chemicals: adenine, hymine, guanine, and cytosine. The DNA molecule consists of hundreds of thousands of atoms. Nevertheless, it is so small that it cannot be seen by the naked eye. Although its weight is no more than one billionth of a gram, this molecule encodes such a vast amount of information that tens or even hundreds of volumes of thick books would be needed to describe it.

The DNA makes its own copy of the molecule, while in every other aspect it acts as the "chief administrator of the living cell" with the help of assistants such as RNA molecules. The DNA molecule is located in the nucleus of the cell, from which it dispatches working RNA molecules for the direct reproduction of protein.[3]

According to Academician A. I. Oparin's theory, primordial proteins and nucleic acids must have originated on the ancient earth *at the same time and independently of one another*, because proteins are necessary for the biological synthesis of nucleic acids, while nucleic acids are necessary for protein synthesis. At the molecular level neither proteins nor nucleic acids are capable of undergoing natural selection all by themselves separately. Only a combination of the two could have undergone natural selection.[4]

If proteins appeared for the benefit of nucleic acids, this definitely means that the purpose of proteins was *determined beforehand*. But by whom? By inanimate matter? But does inanimate matter possess the mind necessary to predetermine the purpose of proteins? If nucleic acids appeared for the benefit of proteins, this also definitely implies that the purpose of nucleic acids was *determined beforehand*. But by whom? By inanimate nature? But does inanimate nature possess the intelligence necessary to predetermine the purpose of nucleic acids?

In the words of the German scientist Roland Glazer, the designs of biological cells are "very well conceived."[5] But by whom? By inanimate nature? But can inanimate nature really "conceive" of anything? No! Inanimate matter does not possess any intellectual or creative capabilities! And proteins, nucleic acids, and the living cell are a product of intellectual creativity. From this we arrive at the scientific conclusion that proteins, nucleic acids, and all biological systems are the product of the creative activity of an intelligent (but not material!) Creator, whom we call God. In contrast to religion, atheism is blind faith in the intelligent foresight of unintelligent matter and in an expedient product of creativity without an intelligent creator.

Essentially, the scientific theory of biological evolution states that all living creatures originated and developed in stages according to a predetermined and expedient program, from lower to higher levels, from inanimate substance to living creature, from the living cell to the human brain. But matter itself merely plays the role of a raw material and blind executor in the process of biological evolution, not the role of an intelligent creator. Matter, which acts as a natural cybernetic system in the process of creative development, quite blindly and absolutely aimlessly carries out the orders of codes which are essentially material copies of an ideal program. God created the ideal program of biological development, and that is why we consider the intelligent God, and not matter, the Creator of all biological elements and systems. After all, you wouldn't call an automatic lathe which produces bolts according to an engineer's program the creator of bolt joints. The intelligence of the engineer created the bolts and the lathe itself. By perfect analogy, God is the Creator of animate and inanimate matter.

Biological evolution as a whole is merely one (the fifth) stage in the evolutionary development of the Universe. At the same time, we can see seven other very significant steps in biological evolution itself:
1) the ideal program of biological evolution created by God in the Ideal

World was converted into energetic codes in the white holes of the Material World.

2) In the clouds of hydrogen plasma, energetic codes were converted into substantial codes at the nuclear or electron level.

3) In the auspicious conditions of the primordial earth, the elementary codes were converted into molecular codes, resulting in the formation of proteins and nucleic acids. This period of biological evolution is commonly known as *chemical evolution*.

4) The protein and nucleic acid molecules combined to form a biological cell, which contains a genetic code passed on by heredity.

5) A certain set of biological cells formed the biological system of plant life.

6) A certain set of biological cells formed the biological system of living fish, birds, and land animals.

7) Then came a human, who is capable of loving and creating consciously, like God. At this point the cycle of biological evolution closed. According to the dialectical law of the negation of the negation, an idea (objective and divine) leads to the formation of an idea (subjective and human) through the medium of matter and material codes. Thus, the formula

$$IDEA \rightarrow MATTER \rightarrow IDEA$$

may be considered proven at both the philosophical level and the level of the natural sciences.[6]

Hence the scientific theory of biological evolution corroborates rather than discredits the Bible. The Bible is opposed not by the scientific theory of biological evolution itself, but by its antiscientific atheistic interpretation, which claims that the expedient evolution of living organisms occurs by itself, spontaneously, without any program, without any predetermined goal, and without any intelligent creator. According to the atheistic interpretation, more advanced organisms originated directly from less advanced organisms by themselves, spontaneously, accidently, or as a result of a nonpredetermined "natural" selection process.

While the term "natural" refers to "nature," the term "selection" has to imply a purpose, because there is no selection without a goal, nor could there be, just as there is no goal without intelligence or will, nor could there be. The presence of a purpose is an essential attribute of any kind of reasonable selection, without which the selection ceases to be a selection. If a process flows without any kind of purpose, then in light of the complete absence of purpose, this process can under no circumstances be called selection. Hence only an intelligent creator could be the inventor of any kind of selection (natural or artificial, natural or engineered), not nature or matter, which possesses no mind whatsoever.

While so-called "natural" selection does exist in biological evolution, the term "natural" merely refers to the practical technique by which the products of creativity are derived, but in no way implies that selection occurs without any goal or without any creator. It can answer the question of what the mechanism of evolution is. But it can never answer the question of who supplied this mechanism with an expedient selection program and why.

Only intelligence is capable of intelligent creation and selection. And creation can only be expedient when the creator knows what he or she needs. But matter is incapable of "knowing" or "wanting." Hence matter could never be its own creator. Saying that "natural selection" was the creator of expedient living structures is tantamount to asserting that the sorting machine was the inventor of sorting.

Even if the hypothesis of natural selection is true, it in no way discredits the Biblical model of the appearance of life and human on the earth. The Bible is not opposed by the hypothesis of natural selection itself, which occurs according to God's ideal program through the medium of genetic codes. The Bible is opposed by atheism's unscientific interpretation of this hypothesis, which tries to claim without any substantiation whatsoever that purposeful and expedient natural selection occurred by itself, without any conscious goal, without any intelligence, and without God. The bankruptcy of this view is obvious, because any expedient selection could be considered natural only in its form and outward manifestation, behind which is concealed a conscious goal. Let us look at an example.

Let us assume that in the morning we left a workshop where a heap of dead parts was lying and in the evening we found these parts assembled in a purposeful way so as to produce a radio. It is undeniably true that the radio was assembled from the parts which were separate in the morning. But different people interpret this undeniable truth in different ways. Some people say that the receiver was assembled from the parts by an assembler (human or machine) according to blueprints drawn up by an engineer. Other people say that the parts combined all by themselves to form the radio, without any assembler, without any blueprints, and without any engineering design.

I am no mystic, and I would find it impossible to believe in an atheist miracle whereby inanimate and unintelligent parts would combine to form an expedient radio by themselves without any assembler and without any engineering whatsoever. Hence I would choose the first interpretation and say without any doubt that the radio was assembled from the parts by an assembler (human or machine) according to an engineer's ideas or blueprints. In this case the weighty and visible automatic assembly machine conceals the weightless and invisible idea of the engineer, without which there would be no parts, no assembly machine, and no radio.

By perfect analogy, objective science asserts that inanimate matter could not be transformed into living creatures by itself and that this sort of evolutionary transformation could only have occurred with the involvement of God and according to God's program. If the expedient selection and purposeful assembly of parts into the needed radio constitute practical proofs of the fact that an intelligent inventor was behind this selection and assembly (and it is absolutely unnecessary for us to see and touch this inventor), the purposeful nature of natural selection and the production of the expedient designs of living systems constitute the best practical proofs that natural selection has an inventor with an extremely high intelligence. We call this inventor God. But atheism, which formally rejects mysticism, in reality believes in it. Atheism blindly believes that purposeful natural selection can occur without any conscious goal and that it did not have any intelligent inventor.

If the term "natural selection" implies the automation of natural processes under the auspices of an ideal program of material development, we should remember that any kind of automation has an intelligent inventor, because selection can only be expedient and purposeful. Hence God, who is extraordinarily intelligent, could only be the inventor of this kind of purposeful selection, not nature, which possesses no mind whatsoever.

If the term "natural selection" means the "survival of the fittest," by which the strong devour the weak, this sort of selection could not promote biological evolution. If "survival of the fittest" truly prevailed in nature instead of an ideal program of material development, then only predators would be left on the earth, because they would have devoured all the people back in the time when people had nothing but rocks in their hands. This sort of purely atheistic principle of "natural selection" is improperly labeled "biological evolution" just as often as enslavers call themselves "armies of liberation." In reality, biological evolution does not mean the assimilation of the weak by the strong, just as liberation does not mean enslavement. From the study of history we know of quite a few examples where the law of "might is right" inevitably culminated in major disasters but did not lead to evolutionary development.

For example, Fascist Germany, in which the "might is right" principle openly reigned supreme, lasted only for 12 years. The most powerful communist regime on the earth, which glossed over its "brute force principles" with flowery slogans, collapsed after 73 years. The United States, which ascribes to more humane purposes, has been in existence for more than 200 years. Any evolutionary development can be a creative and only a creative process, but cannot be the anarchic ("natural") rampage of a "brute force principles." And a creative process would be impossible without an intelligent Creator.

If the term "natural selection" means the natural death of organisms with less perfect genetic codes, and the survival and multiplication of organisms with more perfect genetic codes (and in the opinion of the author, this is exactly the way it is!), then it would be better to call this process "genetic selection," because the basic characteristic of selection is in fact the quality of the genetic code. If the quality of the genetic code were not subject to purposeful selection, then each successive generation of biological organisms would be less viable than the preceding. There would be no evolution or development at all, and life on the earth would have become extinct long time ago, not being able to successfully propagate.

Moreover, the term "natural" is erroneously taken by many people to mean something originating spontaneously, by itself, under the effects of physical forces of an unintelligent nature, with no purpose, no program, no intellect, no creator of any kind. This makes it possible for the atheist to deceive the masses and to remove the element of the sacred.

But no matter what the terms are, the essence of the matter remains unchanged. Whether we call this purposeful selection taking place in physical nature "natural" or "genetic," it pursues a single purpose: perfection of biological systems and organisms. This purpose cannot arise independently of will, and will cannot be independent without intellect. The result of any kind of "natural" selection in the evolutionary process has to be a closed link of the following chain:

**Intellect → Will → Purpose → Program → Code → Selection → Result of Selection**

No successive link can arise, exist or develop without the preceding link. The seventh link is always the concluding stage ("the seventh day") of any evolutionary creation, be it the creation of the entire world, or biological evolution, or the evolution of human relations. In addition, according to the fundamental law of nature, physical selection of biological organisms cannot take place without its nonphysical opposite; that is, without ideal selection and development of intellectual souls, as a result of which the cruel souls of treacherous murderers die out and are removed from circulation once and for all, while the blameless souls are reborn again and again, raising their perfection to a higher level with each succeeding physical reincarnation. According to the law of negation, physical and spiritual selection must follow in a sequence, in turn, one following the other. Were it not for this ideal selection of souls, each succeeding generation on the earth would be more cruel than the preceding. However, history convinces us that the reverse is true: each succeeding generation of people is distinguished from the preceding generation by the goodness of the human soul. From generation to generation, man becomes a relative likeness of the Absolute God to a greater and greater degree. According to the scientific theory of the now-Israeli (formerly, Soviet) biologist Evgenniy Reznitskiy, the purposeful selection along the Absolute God's ideal program occurs not only in the world of living organisms but also in human society.[7]

While all living organisms, ranging from the tiniest microbe to the largest animal, consist of similar or even identical cells, all engineering designs also consist of similar or even identical parts. Hence saying that mammals originated from fish all by themselves, and that fish originated from algae all by themselves is tantamount to saying that carts turned into automobiles all by themselves, automobiles turned into airplanes all by themselves, and airplanes turned into spaceships all by themselves as a result of technological evolution, without any human thought. Technical evolution has in fact progressed from the lowest to the highest levels in the following sequence:

**Cart → Automobile → Airplane → Spaceship**

But the evolution of metal structures does not imply the direct and spontaneous transformation of one kind of vehicle into another, but the evolution of the coherent engineering thought whose program led to the development of new vehicles. By perfect analogy, the presence of similar or even identical cells in all biological systems provides evidence of the coherence of the creative concept of the one God, not that one species of living organism was transformed into another species by itself.

A cherry will never turn into an apple, an apple will never turn into a pear, a pear will never turn into a fish, a fish will never turn into a monkey, and a monkey will never turn into a man. The fossils of ancient animals and plants discovered by archaeologists have convinced us that the evolution of each living species occurred without the slightest traces of the transformation of lower life forms into higher forms.

The hypothesis of the spontaneous transformation of inanimate matter into very simple living organisms and of simple organisms into complex living creatures is tantamount to saying that iron is spontaneously transformed into useful parts, that the parts combine by themselves into the useful automobile, and that a vehicle capable of traveling only on the ground is spontaneously transformed into an airplane capable of flying in the sky. These sorts of transformations of engineering constructions have in fact occurred, but the author of these transformations was human, not metal. Biological evolution, whose author was not inanimate matter or even human, but the absolutely perfect God, has occurred in roughly the same way.

Even Charles Darwin admitted that if the hypothesis of the

metamorphosis of some species of living beings into other ("more refined") species were true, the "number of intermediate links between all species of living organisms would have to have been unimaginably large." But where are these intermediate links? There never were any! Consequently, biological evolution should not be understood as the spontaneous (or "natural") metamorphosis of less refined species of living beings into more refined species. *Biological evolution* is a gradual (nonrevolutionary!) process of the step-by-step formation of different species of plants and living organisms, which are qualitatively different from one another but cannot metamorphose into one another, from inanimate matter ("from the dust of the earth"). Atheism has eagerly tried to instill in naive people the idea that "evolutionary" supposedly means "spontaneous." In fact it is exactly the opposite. The process of biological evolution is not haphazard and spontaneous; it follows an expedient ideal program designed by God and encoded initially at the energetic and elementary levels and then at the level of atoms and molecules. Hence the hypothesis of the "spontaneous" metamorphosis of some species of living creatures into other ("more refined") species is the antiscientific atheistic interpretation of the scientific theory of biological evolution.

As a matter of fact, every living species has developed in an evolutionary manner from a lower to a higher form not spontaneously, but according to its own program, which differs from the programs of all other species and was encoded long ago in the white holes at the photon level. One photon is similar to another photon only in terms of its outward form, i.e., at first glance. But all photons may differ qualitatively from each other in terms of content, because they can carry the energetic codes of completely different ideal programs. Thus, the monkey developed according to its own program, and the human developed according to his own program. Each followed its own evolutionary path from an energetic code to its current condition. So human intelligence has invented and built submarines which plow the seas and oceans at any depth. But who invented the more refined submarine design known as a live fish, whose flexibility, self-control, and maneuverability is still an unattainable dream for the designers of the most advanced ship? There is more wisdom embodied in the fins of a fish than in the design of all the electronic gear on a ship. On the primordial earth there was no fish at all, but then it appeared. Was it nature that created the fish? But does inanimate nature have a will or a mind which would allow it to create better than a human being?

If we admit that inanimate matter has no mind whatsoever, then we must also admit that the fish was created by a nonhuman and nonmaterial intelligence. Without such an ideal otherworldly intelligence, unintelligent nature would have been absolutely incapable of creating such an expedient design as a live fish. If the expedient design of a submarine constitutes practical proof that its designer existed (and we have absolutely no need to see or touch this designer), then the more refined and expedient design of the live fish constitutes the best practical proof of the existence of its nonmaterial but highly intelligent creator whose name is God.

In contrast to religion, atheist "science" is powerless and hence cannot correctly explain the origins of living beings in general and the living fish in particular. Shocked and panicked by the expedient design of a live fish, atheist "science" unjustifiably claims that the live fish was supposedly created by unintelligent nature. Saying that inanimate nature created the live fish means attributing creative intelligence to unintelligent matter. Attributing creative intelligence to unintelligent matter means fetishizing or deifying inanimate matter. The atheistic superstition of the creative abilities of unintelligent nature is one form of antiscientific idolatry, where all matter as a whole is the deified idol instead of some chunk of granite.

We know that the design of a submarine was borrowed from a live fish and is a crude model of it. Atheists attribute this to the fact that scientists learn from nature in their creative activity. But how can an intelligent scientist learn from unintelligent nature? Is inanimate matter somehow wiser than a living human? No! Neither the fish nor matter is wiser than human! What is wiser than human is the nonmaterial intellectual who created the live fish and put wisdom in its design and behavior. Humans learn creativity not from a live fish and not from unintelligent matter, but from their intelligent creator. Nature in general and the live fish in particular are not teachers but merely visual aids which the inventors of submarines used. The teacher is the one who prepared and showed these visual aids. The name of this teacher is God.

The more we study animate and inanimate nature, the more we are convinced that everything in it is arranged extremely efficiently, optimally, intelligently, and sensibly. For example, the conical shape of the human ear ensures the best quality and maximum intensity of sound transmission. Another example would be the goat's horn, which is designed to inflict the sharpest possible blow on an opponent efficiently. Hence its shape ensures maximum strength with minimal weight. Let us note that nature in and of itself does not "know" the subject of the strength of materials and is incapable of performing any kind of strength analysis. One might ask who did perform these analyses.

The atheistic principle of "spontaneous creation," which says that matter is simultaneously the product of its own creativity and its own creator, is absurd and ridiculous, because the *intelligent structure of unintelligent matter provides unmistakable evidence that matter is a product of intelligent creativity, and that a product of intelligent creativity is impossible without an intelligent creator.*

But inanimate matter has no intelligence whatsoever and hence cannot be an intelligent creator. Consequently, we must look for the creator of animate and inanimate matter outside matter, outside the Material World, i.e. in the world of objective ideas, in the Objective Ideal World. Moreover, any creativity definitely implies the necessary presence of will as well as intelligence. Hence the cause of the expedient structure of animate and inanimate matter lies in the creative will of its creator. Religion uses the word God to refer to the Ideal Creator of matter, who possesses an absolutely free will and an absolutely perfect intelligence. God is an absolutely perfect ideal category who created the Material World and is guiding it to perfection. It was God who designed the expedient and purposeful program for the creation and development of matter, gave birth to life in accordance with a definite plan, and has guided its progressive development to a common goal, namely perfection.

The above-cited examples would convince us that living systems are more refined than the machines, devices, and even artworks created by man. There is no doubt that even the most refined cybernetic model of the tiny ant would be larger in size than the Eiffel Tower. The wisdom by which a variety of problems were solved in the design of living beings has dazzled the human imagination. A special science known as *bionics* has been created for the purpose of studying the intelligent solutions of different creative problems used in biological systems for the purpose of applying them to engineering. Animate nature is an abundant source of knowledge for the development of bionics. The designs of many splendid architectural structures were borrowed from plants, aircraft designs were borrowed from birds, ship designs were borrowed from fish, and so forth.

The high quality of the expedient designs found in animate and inanimate nature continues to remain an unattainable dream for the designers who design the most advanced machines or devices. The expedience of material structures has been proven so convincingly by the natural sciences that it has already ceased to be a subject of debate. This expedience has been unconditionally acknowledged by both religion and atheism. Religion and atheism diverge on the question of who was the creator of the expedient structures of animate and inanimate matter. Was it unintelligent matter or the intelligent God? Atheism unjustifiably holds the first opinion, while religion has scientific proof of the second opinion.

For example, Viktor Pekelis, a scholarly atheist, has answered this question in the following way: "We often compare the devices we have created with living beings and see how far behind nature we are in our creativity."[8]  If living beings were created by nature, then who is responsible for striking the optimal balance between the birth rates of males and females? "Why do animals gather in packs and herds and birds in flocks? Who assembled the gigantic feathered congregations? What force has organized the bird bazaars? Who is the air traffic controller who controls the flights of bird airlines over millions of kilometers?"[9]

Does inanimate nature really have the will, the mind, or the intelligence that would allow it to create better than humans can? If we admit that inanimate nature has no mind whatsoever, then we must also admit that expedient living organisms were created by a nonmaterial intelligence. Unintelligent nature would never have been able to create such an expedient structure of living creatures without this kind of otherworldly intelligence. If the expedient design of a machine or mechanism constitutes proof that its designer existed (and we have absolutely no need to see or touch this designer), then the more refined and expedient structure of animate and inanimate matter constitutes the best practical proof of the existence of its nonmaterial but highly intelligent creator, whose name is God.

In his final book, the Soviet academician Oparin sums up the sad results of atheism's ignoble battle against religion in the following way:

> By trying to conceive of the living being as some sort of complex mechanism or piece of machinery and studying the individual substances of the organism, its structure, and the processes which take place in it, mechanistic materialism has accomplished quite a bit in understanding life. But it has invariably been completely powerless to solve the problem of the origins of life. The analogy between an organism and a machine was completely fruitless in this

respect. After all, a machine does not appear by itself in the inorganic world. It is a product of the creative efforts of its inventor and designer. Hence this analogy has inevitably led even materialistic-minded scientists to profoundly idealistic conclusions and to the realization that life-giving intelligence is necessary for the origins of life.[10]

The atheistic notion of the impossibility of the existence of a highly organized objective idea without matter is a retreat from the natural sciences and from science itself, which have convincingly proven that unintelligent matter is the expedient product of an intelligent idea.

On the basis of the most recent information from the natural sciences, we may concisely summarize the *scientific model of biological evolution* as follows:

*And God created an ideal program of biological evolution by which life would inevitably and logically develop, at first in the water and then on land. The elements of this program created by God in the Ideal World would be converted into energetic codes ("seeds of biological life") in the white holes of the Material World. These energetic codes would be converted into substantial codes at the nuclear or electron level in the clouds of hydrogen plasma. Under the auspicious conditions of the primordial earth, these elementary codes would be converted into codes at the molecular level, accompanied by the formation of proteins and nucleic acids. The protein and nucleic acid molecules would combine to form a living cell, which contain hereditary genetic code. A certain set of living cells would form a biological system such as a plant, fish, bird, or land animal.*

*"And God said, Let the waters bring forth abundantly the moving creature that hath life, and fowl that may fly above the earth in the open firmament of heaven."[11]*

*And God first created living creatures in the water, and then the large fish and sea animals, and then God created winged birds which fly through the air. Every species of biological system develops in an evolutionary way from the lowest to the highest form according to its own program, which differs from the program of all the other species and was encoded at the energetic or elementary level. Initially each species of living organism arose from its own "seed," which differs from the "seeds" of all the other species. Later, all living creatures were born, developed, matured, produced offspring after their kind, aged, died, and were born again. The Ideal God created living creatures not with material hands or a magic wand but by means of an ideal program and genetic codes, which operated as natural cybernetic systems.*

"And God created great whales, and every living creature that moves, which the waters brought forth abundantly, after their kind, and every winged fowl after its kind."[12]

*And God was satisfied with the results of his creation, because the biological systems contained all the conditions necessary for the next phase of the evolutionary development of the Universe, namely the phase of intelligent evolution.*

"And God saw that it was good."[13]

*And God worked so that the fish and the birds were fruitful and multiplied and filled the water in the seas and the space over the earth.*

"And God blessed them, saying, Be fruitful, and multiply, and fill the waters in the seas, and let fowl multiply in the earth."[14]

*And the fish arose in the water ("and there was evening"), and the birds flew up into the sky ("and there was morning"): thus went the fifth stage of the creation of the Universe by the absolutely perfect God. We call this period of the creation of the Universe the stage of biological evolution: ("fifth day").*

So what can atheism use to counter this scientific model of biological evolution besides its own whims and illusions? It can only counter the scientific model with its own totally unproven hypotheses that long ago inanimate substance turned into living beings all by itself, without the involvement of any kind of intelligent creator. But then we might appropriately ask where the scientific proofs of atheism are. And the answer is quite simple: "scientific" atheism has no scientific proofs nor could it have any!

It only has its convenient initial hypotheses which hundreds of millions of people have to believe blindly, even though it is logically impossible to believe them, because the natural sciences have definitely established that the spontaneous origin of life is altogether impossible. Back in 1864 the French Academy of Sciences awarded a prize to the French scientist Louis Pasteur (1822-1895) for his experiments which convincingly refuted the unscientific notions of the spontaneous generation of living organisms from inanimate substance. Louis Pasteur himself said that "The theory of the spontaneous generation of life will never rise again after this fatal blow."

Academician A. I. Oparin has written the following on this subject:

> This knocked the ground from under the feet of those researchers who saw sudden spontaneous generation as the only solution for the problem of the origin of life. Deprived of the opportunity to experiment in this direction, they fell into a deep depression and basically quit trying to solve this "accursed" problem and only looked for different justifications for its insolubility. This was the soil from which sprung, in Bernal's phrase "the practice of subterfuges," which were merely designed to evade a rational solution to the problem of the origins of life without essentially explaining anything.
>
> One such "subterfuge" was the hypothesis which stated that the origin of the first living being on the earth was not a logically predetermined event but an extremely rare 'fortunate accident.'[15]

Corresponding Member of the Soviet Academy of Sciences Joseph Shklovsky has objected to this sort of "fortunate accident" in the following way: "To think that such a complex piece of machinery as 'proto DNA' and the protein enzymes necessary for its operation could be produced by pure happenstance, by the 'jumbling' of individual multiatomic molecular building blocks, is equal to believing in miracles.

It would be much more probable to assume that a monkey pounding away at random on the keyboard of a typewriter could produce Shakespeare's 66th sonnet by accident."[16]

Above we have seen a purely scientific model of the fifth stage of the evolutionary development of the Universe. How does it differ from the Biblical model? Analysis and comparison will convince us that there is no essential difference between the semantic contents of the scientific and Biblical models. The only difference lies in the form and style of presentation and in the distinctive features of the ancient and modern languages. For example, while in the Biblical model God created the fish and the birds by means of the Word ("And God said"), in the scientific model biological systems are generated by means of the ideal program created by God. In both cases we are talking about the will of God. The simple people of antiquity had no conception of modern cybernetic systems and hence the Biblical expression "And God said" should be translated into modern scientific language as follows: "And God created an ideal program."

The Ideal God does not have any material hands, feet, beard, and so forth. Hence God creates living beings not with his hands or with a magic wand, but by means of the "Divine Word," i.e., by means of an ideal program and its material codes. It is atheism (not the Bible) which depicts God as a material old man with a long beard and a magic wand in order to simplify its war against the Bible. The Bible, however, depicts God as an ideal (not a material) category. Thus, modern science corroborates, and does not discredit, the Bible and refutes atheism, not religion.

"Scientific" atheism is scientific only with respect to its form, its cover, and its title. But in essence and with respect to its real content, atheism is blind antiscientific faith in an expedient product of creation without an intelligent creator. Drawings without an artist, inventions without an inventor, a machine without a designer, a program without a programmer, and laws without a legislator are all unscientific fairy tales which not even the most backward human being could believe. But, under the onslaught of the overpowering propaganda of atheist leaders and with the silent complicity of all humanity, hundreds of millions of people have been forced to believe blindly in the spontaneous creation of matter without an ideal God and in the material product of creativity without an intelligent creator, even though it is logically impossible to believe in it.

In the depths of the earth and the sun a program was encoded by which the fifth stage of the evolution of the Universe would be accompanied by the emergence of life. Compare this scientific principle with the following passage from the Bible: "And God said, Let the waters bring forth abundantly the moving creature that has life."[17] The meaning of the existence of the earth and the sun primarily lay in the fact that biological evolution would have been impossible without them. The initial condition for the fifth stage of the evolution of the Universe was the existence of the earth and the sun, and its inevitable end result was the appearance of living creatures. But why was it necessary for living creatures to appear?

The "Fifth Day" (Morning)

The natural sciences have authenticated the fact that the organisms of the fish and birds were built from biological cells earlier than were the organisms of land animals, as is stated in the Bible. ("And God created great whales, and every living thing, which living creature that moves, which the water brought forth abundantly. . . fifth day.")

The "Sixth Day" (Evening)

The natural sciences have authenticated the fact that the organisms of land animals were built from biological cells later than the organisms of the fish and the birds, as is stated in the Bible. ("And God made the beast of the earth after its kind, and the cattle after their kind, and anything that creeps upon the earth after its kind. . . And it was the evening of the sixth day.")

# =15=

## THE SIXTH STAGE OF CREATION: INTELLIGENT EVOLUTION: ANIMALS AND MAN

**T**he Biblical Model in Ancient Hebrew

And God said, Let the earth bring forth the living creature after its kind, cattle, and creeping thing, and beast of the earth after its kind: and it was so. And God made the beast of the earth after its kind, and cattle after their kind, and every thing that creeps upon the earth after its kind: and God saw that it was good. And God said, Let us make a human being in our image, after our likeness: and let them have dominion over the fish of the sea, and over the fowl of the air, and over the cattle, and over all the earth, and over every creeping thing that creeps upon the earth.

So God created a human being in God's own image, in the image of God were they created, male and female. And God blessed them, and God said unto them: Be fruitful, and multiply, and replenish the earth, and subdue it, and have dominion over the fish of the sea, and over the fowl of the air, and over every living thing that moveth upon the earth. And God said, Behold, I have given you every herb bearing seed, which is upon the face of all the earth, and every tree, in which is the fruit of a tree yielding seed; to you it shall be for to eat. And to every beast of the earth, and to every fowl of the air, and to every thing that creepeth upon the earth, wherein there is life, I have given every green herb for to eat, and it was so. And God saw every thing that God had made, and, behold, it was very good. And there was evening and there was morning, sixth day.[1]

### The Biblical Model in Modern Popular Scientific Language

And God said: let the different species of living creatures multiply on the earth: and creeping things after their own kind, cattle after their own kind, and the beasts of the earth after their kind. And it was so. And God created the beasts of the earth after their own kind, cattle after their own kind, and all the creeping things of the earth after their own kind. All of them were born, developed, matured, produced offspring after their own kind, aged, died, and were born again. And God was satisfied with the results of this creation, because God had created all the conditions necessary for the appearance of a special living creature on the earth, namely a human being, who possesses high intelligence and creative abilities. And God said: let us create a human being after our own image, after our likeness, and let them have dominion over the fish of the sea, and over the fowl of the air, and over the cattle, and over all the earth, and over every creeping thing that creeps upon the earth.

And God created a human being after God's own image and likeness. And humans appeared on the earth as the relative likeness of the absolute Creator, as the material image of the nonmaterial God, and as the inseparable unity of sexual opposites, male and female. And God blessed them and said to them: be fruitful and multiply and replenish the earth and subdue it, and have dominion over the fish of the sea, and over the fowl of the air, and over every living thing that moves upon the earth.

And God said: I, God, have created the primal seeds of different plant species from which shall grow seed-bearing herbs and fruit trees which shall multiply over the entire earth from year to year. These plants, the herbs, vegetables, and fruits of the trees shall be plant food for humans. Other species of animals shall also live on plant food: all the animals of the earth, all the birds in the sky, and all the creeping animals. And it was so. And different species of animals arose on the earth ("and there was evening"), and humans arose ("and there was morning"): thus went the sixth stage of the creation of the Universe by the absolutely perfect God. We call this period of the creation of the

Universe the stage of *intelligent evolution:* the sixth day. And God was satisfied with the results of all of this creation, because in the person of man and woman God had created worthy helpers who was capable (like God) of creating purposefully and consciously. ("And God saw every thing that God had made, and, behold, it was very good.")

## The Modern Scientific Model

And God created an ideal program of biological evolution by which living creatures would inevitably and logically appear in both the water and on land. According to this program, a special living creature who possesses high intelligence and creative abilities, would arise on the earth along with different species of animals. Hence the stage of intelligent evolution is not only a continuation but a part of biological evolution. And the intelligent evolution of man and woman, which was rooted in biological evolution, is still continuing and will continue for a very long time to come.

The mechanism by which humans appeared on the earth has still not been scientifically established. We would like to draw the reader's attention to a hypothesis which states that we were created by God in the following six stages:

1) The ideal program for the creation of human beings, which was designed by God in another (nonmaterial) world, was converted into energy codes at the photon level.

2) The energy codes were converted into substantial codes at the elementary level. These elementary codes constituted a special primal seed for the human of the future.

3) Under the auspicious conditions of the ancient earth, these primal seeds were converted into codes at the molecular level, accompanied by the formation of proteins and nucleic acids.

4) The protein and nucleic acid molecules combined to form a primal human biological cell which contained hereditary genetic code of a

special kind. These biological cells are no longer present on the earth. If this were not true, we would still be witnessing the nonsexual birth of the primal human on the earth from a substantial seed.

5) In an era limited to a relatively short period of time, in all likelihood no more than 200 million years ago and no later than 50,000 years ago, the weightless and bodiless photons of sunrays carried the energy codes of an ideal program from another world to the material earth. According to the program, these biological cells of a special kind would inevitably and logically form the original human "flesh" or "embryo." Ultimately a biological system was formed which could not as yet be called human but which already contained an encoded program for its own inevitable metamorphosis into a human being. This "embryo" of the original human, like an egg, did not have any sex or consciousness and hence the Bible calls it "the human being (Adam) upon whom God caused a deep sleep to fall."[2] In Genesis 2:22 the Bible calls the main portion of this "embryo" the "rib" in the same way that we call the main (most important for the genesis of life) part of an egg the "yolk."

6) Subsequently this "embryo" divided into two sexual halves in the same way that one cell divides into two ("And the rib, which God had taken from Adam, was made into woman, and she was brought unto man"). In Egyptian hieroglyphics the rib is a symbol for woman. A man was formed from one half of this embryo (called Adam), while a woman was formed from the other half (called Hava, or Eve).

In 1992 many English-language newspapers and magazines published articles which reported that biologists had definitely established that all people on the earth were descended from one and the same protomother and one and the same protofather.

As we already know, God created the fundamental law of nature which states that nothing material can arise without its opposite. In accordance with this law, God split the single embryo of the primal human into two sexual opposites: a man and a woman. And God formed a true couple from a single flesh so that male and female would be passionately attracted to one another. And as the highest reward for their refinement would the true spouses find each other, become husband and wife, fuse into one, and acquire true happiness. The English word "husband" originated from two ancient words, "hus" (enjoyment) and "band" (unity), i.e., "union for pleasure." "And Adam said, "This is now bone of my bones, and flesh of my flesh: she shall be called Woman, because she was taken out of Man."[3]

So "the wife was taken from the rib of the husband." Thus appeared primal humans, who, like God, were capable of consciously loving and creating. This is where the process of the creation of the original human beings by God concluded and the cycle of biological evolution closed. According to the dialectic law of the negation of the negation, an idea (objective and divine) leads to the formation of an idea (subjective and human) by way of matter and material codes.

Subsequent human multiplication and development would occur sexually by means of hereditary genetic codes. This kind of human "sexual multiplication" and further intelligent development would require minimum supervision on God's part. Hence it could be called the seventh (culminating!) stage of both biological evolution and the evolutionary development of the Universe as a whole, when God would get a well-deserved rest from all this creative activity.

The natural sciences have definitely established that biological cells are the "building blocks" of all biological systems, which include all living organisms, ranging from the simplest bacteria to the largest animals which have appeared on the earth in different eras. The body of the living person, who possesses high intelligence, consists of biological cells in the same way as plant organisms, which possess no mind whatsoever, not even the most primitive mind. The only difference between them is in the semantic content of their genetic codes, which carry ideal information and explicit commands from another world.

The use of completely identical parts and assemblies in different engineering designs is commonly known as *unification* (or standardization). Unification makes it much easier to automate assembly processes. Hence it is highly expedient and it is the sacred duty of any designer. In the old Soviet Union a designer could be put on trial for failing to perform this duty.

If similar or even identical parts can be used in an automobile and a spaceship, this in no way implies that in the process of technological evolution the automobile turned into a spaceship without any engineering ideas or without any intelligent creator. By perfect analogy, if the bodies of plants, monkeys, and humans consist of biological cells, this in no way implies that in the process of biological evolution a monkey originated from a plant and that a human originated from a monkey without any creative ideas or without any intelligent creator. On the contrary, the standardization of the basic components of biological cells and living organisms provides convincing evidence that the evolution of inanimate substance into living beings and so called "natural" selection could not have occurred by themselves without the involvement of a supreme divine intelligence.

But then we might quite naturally ask how and why single cells combined or were transformed into multicellular plant organisms in some cases and into the extremely complex organisms of living creatures or human beings in other cases.

Every biological cell contains a genetic code, which is a material replica of an ideal (genetic) program. This ideal program, which was created by God in another world, provides each biological cell with complete instructions on how to behave in any situation by means of its genetic code. Each biological cell blindly and unswervingly carries out any command it receives from God via its genetic codes and via the laws of nature whose general compilation constitutes the ideal program of biological evolution. This means that the ideal program created by God completely defines the standard of behavior for each biological cell in any situation. Some biological cells are told to combine into multicellular plant organisms, while others are told to combine into the bodies of a shark or eagle, depending on the environment. In the process the environment and the situation also emerge and change in response to other programs which were also created by God.

The combination of programs creates the conditions by which different living creatures inevitably and logically form from solitary living cells and are never transformed into one another. Fish emerged in the sea, land animals emerged on land, and amphibians emerged in small bodies of water which often dried out and were then refilled with water. While the process by which this transformation of inanimate substance into living creatures is commonly known as "natural" selection, this in no way implies that sea and land animals can be transformed into one another as a result of "natural" selection. We can quite confidently say that if we were to dry out the Atlantic Ocean, none of the billions of fish there would turn into a sheep or a camel. We can also be sure that no human being could ever turn into a fish, even if the entire globe were covered with a thick layer of water.

If natural selection does exist in biological evolution, then it ends with the formation of different species. Hence biological species is the name we used for that category of living creatures whose breeding yields fertile offspring.

Interspecies breeding (such as breeding a lioness with a camel or a man with a monkey) does not produce any offspring. The monkey emerged according to one program, while man emerged according to a completely different program. While the first program provided for the emergence of an agile animal, the goal of the second one was to produce a living being with high intelligence and creative abilities.

If atheism were correct and biological evolution were the spontaneous metamorphosis of some (lower) species of living beings into other (higher) species, the ancient mammals who existed more than 200 million years ago would have been transformed long ago into the most intelligent creatures on the earth, while humans, who only appeared 60,000 years ago, would be the most stupid creatures on the earth. But the scientific facts indicate the opposite: "Human beings has traveled a relatively short path of evolution, but in terms of complexity and mental capacity stands taller than the animals."[4] This was undoubtedly facilitated by the features of the ideal program for the creation of man designed by God. The special rules of this program accelerated the evolutionary development of humans and their central nervous system. Thus, the monkey has absolutely no relation to this exclusive program for the creation of the human species by God.

Man and woman, of course, could have appeared first on another planet in another solar system. After all, the primal seeds of human life could have fallen on other planets as well as the earth. Adam and Eve, from whom the entire human race on the earth today descended, could have landed from any other planet onto our earth.

The equivalence of the Biblical word "dry land" in this context to the modern scientific term "planet" provides evidence that life could have developed not only on our planet but on any other planet. Consequently, we cannot rule out the possibility that human life may have made its first appearance on another planet and quite recently moved to the earth from there. This kind of interplanetary migration would have required highly advanced technology. But then we might logically ask how a human migrant could have "forgotten" such technological achievements. It's quite simple. First of all, it could have occurred as a result of a disaster such as the "global flood." Secondly, imagine that you and your wife (or husband), who have a good knowledge of advanced technology, move to another (unknown) planet. Despite your profound knowledge of advanced technology, you cannot build power plants, airplanes, or rockets on the new planet, because you would need specialists in many fields of technology, and one married couple would find it physically impossible to combine such a large number of occupations. Your children and grandchildren would gradually forget about airplanes and rockets until their distant descendants arrived at technological advances on their own. That is what might have happened with Adam and Eve.

But this scenario in no way implies, however, that man and woman could have undergone the first part of their evolutionary development on one planet and the second part on another planet. Human life must have traveled the entire path of evolutionary development from a primal seed to a living being with high intelligence and creative abilities on the same planet. Otherwise, it would not have been able to move from one planet to another on its own. No matter which planet humans first appeared on, they could only have emerged as a product of intelligent creativity and, thus, could only have been created by God.

There are very strong scientific arguments against human settlement on the earth from another planet. First of all, no signs of biological life, let alone human, have been detected on any other planet in our solar system besides the earth. Secondly, humans would have had to fly at least 4 terrestrial years from any other planet (even the closest to us), even if they had flown at the speed of light, which seems practically impossible.[5]

Modern science has been occupied with the problem of creating self-developing and self-multiplying cybernetic systems. In the process it has been proven that a bisexual multiplication system is the best. Unisexual systems degenerate (and do not develop) because cumulative error as well as genetic codes are passed on from generation to generation. Multisexual systems would develop worse than bisexual systems, because of the lower probability of many sexes combining to breed. From this we may conclude that living creatures are bisexual not just by chance but in accordance with a predetermined goal and as a result of an intelligent (and conscious) solution of the problem. Unconscious natural selection could only be the outward form or physical manifestation of a conscious decision. And a conscious decision could only be made by an ideal intelligence. That is why human life appeared on our earth in the biological form of sexual opposites: man and woman.

Men and women cannot exist and multiply without each other. They presuppose each other. They exist for each other. If man and woman emerged for the sake each other, then this definitely means that human purpose was determined in advance. But by whom? By inanimate matter? Does inanimate matter really have a mind necessary to predetermine the purpose of a human life? Inanimate matter has no intellectual or creative abilities! From this we may arrive at the scientific conclusion that men, women, and all living creatures are the product of the creativity of the intelligent creator whom we call God. Atheism, on the other hand, constitutes a blind faith in the spontaneous metamorphosis of inanimate and unintelligent matter into animate and thinking matter.

Cybernetic systems have been invented by human intelligence to perform "highly skilled" work. The perfection of the living cell is the limit toward which the perfection of cybernetic systems will always strive in the process of technological development but will never achieve. The human brain is a much more complex combination of billions of living cells. But who invented the human brain, which works better, more economically, and more reliably than the most advanced cybernetic system?

The high quality and reliability of the human brain with its minimal size and weight continues to be an unattainable dream for the designers who produce the most advanced cybernetic systems. Each living cell of the cerebrum contains more wisdom than an entire computer. Did nature really create the living brain of a human being? And does inanimate nature really have the will, mind, or intelligence to create such a complex and highly organized system as the human brain? If we admit that inanimate matter has no mind whatsoever, then we must also admit that the living brain of a human being was created by a special nonhuman intelligence which possesses no material attributes. Unintelligent matter could never have created the intelligent design of a living human brain without this kind of otherworldly intelligence.

If the expedient design of a cybernetic system provides practical proof that its intelligent designer existed (and there is no need for us to see this designer with our eyes or touch him with our hands), then the more refined and more expedient design of the human brain constitutes the best practical proof of the existence of its nonmaterial but highly intelligent creator, whose name is God. Academician Kapitsa commented as follows on the subject: "Bionics is often called a young science. This is not true. After all, God was engaged in bionics when God created people "after God's own image and likeness."[6]

If atheism calls an idea or intelligence a property of "highly organized" matter, then in the process it is tacitly implying that even inanimate matter possesses intelligence. Otherwise, the primordial "loosely organized" matter could never have transformed itself from simple lifeless atoms into the "highly organized" human brain.

Atheist "science," which has been amazed by the expedience of the human brain's design and is unable to provide a logical explanation for it, has quite groundlessly asserted that the human brain was supposedly created by inanimate nature. To say that the living brain was created by inanimate nature means attributing creative intelligence to unintelligent matter. Attributing creative intelligence to unintelligent matter means fetishizing or deifying inanimate matter. Atheistic superstition, according to which the creator of the biological brain and nonmaterial intelligence is supposedly unintelligent nature, is in essence one of the forms of antiscientific idolatry, where in the role of the deified idol you have no wooden knick-knack but an unintelligent nature in its entirety.

On the basis of extant information from the natural sciences, we may summarize the scientific model of intelligent evolution as follows:

*And God created an ideal program of biological evolution by which living creatures would inevitably and logically appear on land as well as in the water. And it was so. "And God said, Let the earth bring forth the living creature after its kind, cattle, and creeping things, and the beast of the earth after its kind: and it was so."[7]*

And different species of living creatures appeared on land as well as in the water. They appeared as a result of the evolutionary development of the "seeds of life," which contained the material codes of an ideal program created by God at the nuclear or even at the photon level. Originally, each living species formed from its own seed, which differed from the seed of any other species. Subsequently each living species has multiplied after its own kind by sexual means.

Interspecies breeding (such as breeding a wolf with a horse or a man with a monkey) produces no results. Each individual living creature is born, develops, matures, produces offspring after its own kind, ages, dies, and is reborn. The ideal God created primal living creatures not with material hands or a magic wand, but by means of an ideal program and genetic codes which operated as natural cybernetic systems. This ideal and perfect God prepared all the necessary conditions for the appearance of a special living creature on the earth, namely a human being, who possesses high intelligence and creative abilities. *"And God made the beast of the earth after its kind, and cattle after their kind, and every thing that creeps upon the earth after its kind: and God saw that it was good."[8]*

And God created an ideal program of intelligent evolution by which the biological reflection of the ideal God would logically and inevitably appear. In contrast to the ideal God, human beings must have a material body, because their purpose is to assist God's creative activity on earth. That is why humans (like God) must possess high intelligence, and this high intelligence will give them power over the entire earth: over the fish in the sea, the birds in the sky, and over all the living creatures inhabiting the land. *"And God said: let us create a human being after our own image, after our likeness, and let them have dominion over the fish of the sea, and over the fowl of the air, and over the cattle, and over all the earth, and over every creeping thing that creeps upon the earth."[9]*

And this being appeared on the earth as the inseparable unity of sexual opposites (man and woman), as the material image of a nonmaterial intellect, as the relative likeness of the Absolute God. God's absolutely perfect intelligence is the limit toward which human intelligence will always strive in its evolutionary development but will never reach. The absolute God created human beings not with material hands or a magic wand but with the "Word of God," by means of the genetic codes and natural laws which constitute an ideal program for intelligent evolution. "So God created Adam in God's own image, in the image of God was Adam created; male and female were they created."[10]

And God blessed them by endowing them with intelligence and will, and God gave this will a certain amount of freedom so man and woman could help God in the creative process and consciously strive for perfection. And God endowed them with intelligence and with the creative abilities to the extent that they would achieve power over all the animals of the earth, the fish in the sea, and the birds in the sky. "And God blessed them and said to them: Be fruitful and multiply and replenish the earth and subdue it, and have dominion over the fish of the sea, and over the fowl of the air, and over every living thing that moveth upon the earth."[11]

According to the ideal program of biological evolution created by God, the different plant species, seed-bearing herbs, and fruit-bearing trees would multiply over the entire earth from year to year. These plants, herbs, vegetables, and fruits of the trees would become plant food for sustaining human life. "And God said, Behold, I have given you every herb bearing seed, which is upon the face of all the earth, and every tree, in which is the fruit of a tree yielding seed; to you it shall be for to eat."[12]

Other species of living creatures would also live off of plant food, all the animals of the earth, the birds in the sky, and all the creeping animals. And it was so. "And to every beast of the earth, and to every fowl of the air, and to every thing that creeps upon the earth, wherein there is life, I have given every green herb for to eat, and it was so."[13]

And different species of animals arose on the earth ("and there was evening") and humans, who possesses intelligence, arose on the earth ("and *there was morning"): thus went the sixth stage of the creation of the Universe by the absolutely perfect God. We call this period of the creation of the Universe the stage of intelligentevolution: "sixth day."*

*And God was satisfied with the results of this creation, because in the person of man and woman God had created a worthy helper capable (like God) of purposeful and conscious creation. And humankind would have to justify the high trust which God had placed in it. "And God saw everything that God had made, and, behold, it was very good."*

"Scientific" atheism has embraced the theory according to which human beings descended from apes. What basis does atheism have for this theory? First of all, it asserts that the apes appeared on the earth before human life did. Secondly, the human body is very similar to a ape's body. By following this extremely atheistic logic, we could prove any absurdity. For example, by analogy we could say that Yemelyan Yaroslavsky, the father of scientific atheism, came from Boris Godunov.

And as a matter of fact, Boris Godunov lived before Yemelyan Yaroslavsky. Secondly, there is a greater similarity between the biological structures of Yemelyan Yaroslavsky and Boris Godunov than between the structures of apes and men. Hence there is more truth in the obviously false hypothesis that Yemelyan Yaroslavsky originated from Boris Godunov than there is for the "hypothesis" that men evolved from monkeys.

The only thing left is to be amazed at is how "scientific" atheists, approvingly nodding their wise heads, could calmly swallow such an unscientific pill as the hypothesis that a human life evolved from apes. *The hypothesis of man's evolution from apes is refuted by science in a very simple manner: men and apes belong to different biological species, and their crossing does not produce progeny.*

The Bible is not contradicted by objective science, which has studied the process of the programmed evolution of every species of biological system from the simple to the complex, from the simplest living cells to the living human body. The Bible is opposed by the unscientific atheistic interpretation of this science by which apes spontaneously evolved into humans. To say that man evolved from an ape is tantamount to saying that an airplane spontaneously turned into a spaceship without any designer. By perfect analogy we can say that the similarity between certain organs of the ape and humans merely provides evidence of the coherence of their creator. The appearance of the human species, who is more complex and perfect than an ape, merely indicates the evolution of the creative activity of their creator. But this in no way indicates that man evolved from an ape or that a monkey at some stage in its evolution supposedly became human all by itself. It is perfectly clear that monkeys developed according to its program, and that humans developed according to theirs.

To the above we should add that the hypothesis that humankind evolved from apes does not let atheism off the hook, even though atheism has grasped it as firmly as a drowning man grasps at straws. Even if we had evolved from apes, this would in no way mean that a humanlike ape was spontaneously transformed into a human being, since the natural sciences have definitely proven that any development of the biological form of a living creature follows an ideal program encoded in living cells. And the author of this program could only be God, who possesses a supreme intelligence, not nature, which possesses no mind whatsoever. In any case, the creator of the human species was God, not nature!

Above we have given a purely scientific model of the sixth stage of the evolutionary development of the Universe. So how does it differ from the Biblical model? Analysis and comparison will convince us that the scientific and Biblical models are basically the same in terms of their essence and semantic content. The only difference lies in the form and manner of presentation and in the distinctive features of the ancient and modern languages. The scientific model corroborates the Bible and discredits atheism.

That is why atheism has desperately attempted to turn the power-lessness of its ideology into slander not only against the Bible but against religion. Atheism has not only slandered the Bible, it has unpardonably distorted the contents of the Bible in the eyes of the public. It has fabricated lies, attributed these lies to the Bible, criticized its own lies after calling them Biblical, and has used this as the basis for stating that science discredits the Bible. For example, Yemelyan Yaroslavsky wrote that "God has a beard, a mustache, two eyes, two ears, a nose, two arms, two legs, just like a man."[14]

But this portrait of God is utter slander against the Bible, because the Bible contains nothing of the kind. It was the atheists who conceived of God as a bearded old man who created man with his hands or a magic wand in order to make their ignoble war against the Bible easier. According to the Bible, God is an absolutely perfect ideal category with no material attributes whatsoever. Hence, we should look for the similarity between man and God not in a beard or a mustache, but in man's spiritual and mental qualities: like God, we are capable of loving and thinking. We are not the descendants of apes but rather the biological image and relatively perfect likeness of the ideal and absolutely perfect God. The ideal God created living creatures (including intelligent life) not with humanlike hands or a magic wand, but by means of the "Word of God," i.e. by means of an ideal program, which was initially encoded at the level of photons and elementary particles and then in biological cells.

We are the material image of the immaterial God, the relative likeness of the Absolute Creator. We are expected to love and create like God. Herein lies our primary essence. If one is full of hatred and destruction, then one becomes the opposite of God, and not God's likeness. Therefore, such a person ceases to be human.

But the less we err, the less we hate. The less we hate, the more we build and create. The more we create, the more Godlike we become. Hence only a person who is free of the heavy burden of error and hate and who can love and create like God, who created everything beautiful and who created us, can be a true helper to God. That is why the conscientious search for the objective truth is not a trifling matter but rather the sacred duty of every person who wishes to achieve true happiness.

The global system of living creatures contains an encoded program by which the sixth stage of the evolutionary development of the Universe was accompanied by the appearance of human life on the earth and the development of human intelligence. Let us compare this scientific principle with the following passage from the Bible: "And God said, Let us make a person in our image, after our likeness."[15] The meaning of the existence of a living cell lies in the fact that the human brain could not exist without it. The starting point for the sixth stage of the evolution of the Universe was the living cell, and its end result was the human brain. But for whom and why was human intelligence in the Universe needed? A separate book should be devoted to this subject.

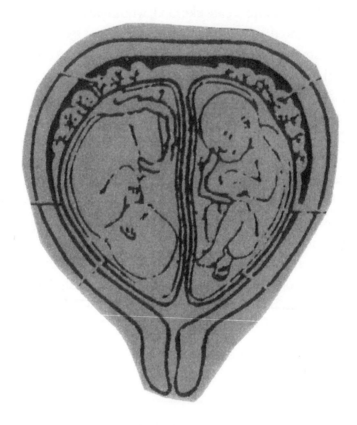

The "Six Day"

And God created a program of intellectual evolution, in accordance with which man originated on Earth as a material image of the non-material God, and as a relative likeness of the Absolute Creator. ("And God created man in God's own image. . . And it was the morning of the sixth day."

Adam and Eve by Jan Van Scorel, 1540

# =16=

## CONCLUSION

On the basis of scientific information, we have demonstrated that the Biblical stories of the origins of the Universe and of life on our planet, culminating in human life, are correct. By some inexplicable intuition, the Biblical author, living over 3000 years ago, had grasped scientific truth. The essence of this truth is that there can be no creation without a creator. But this truth was also grasped by many other ancient cultures around the world. From the ancient Babylonians to the ancient Maya, from tribes living in the Polynesian islands to tribes living in the North Pole, nearly every human community has left us a story or a myth of creation. Invariably, those stories tell of several gods involved in the process of creation, rather than the one absolute and transcendental Biblical God. But there is another fundamental difference between the Biblical stories of creation and all the others. The difference can be found in the words "And God saw that it was good." By repeating these words at every stage of creation, the Bible is making a value judgment which is not found in any other ancient creation story, not even in Greek mythology. Whereas in other stories of creation the gods create the world essentially to amuse themselves, the Hebrew God creates the world or the Universe for what we might call a moral purpose (good, rather than evil). This moral purpose is not a human invention. It is part of the natural law. It is enunciated in the Biblical Laws of Moses, most importantly in the Ten Commandments, and culminates in the Judeo-Christian belief in the ultimate redemption of humankind, either through a human emissary known as the Messiah, or the coming of a messianic age.

We shall first summarize *the scientific model of the creation and* stage *evolution of the Universe*, and then conclude with some rumination about the messianic idea.

**The First Day**

There was nothing: neither space, nor time, nor anything that would have to exist and develop in space and time. There was no kind of world or Universe; neither light, nor darkness; neither galaxies, nor stars; neither the sun nor the earth; no substance, no energy, neither molecules nor atoms; neither nucleons nor electrons. There was rather but a single and absolutely perfect God, existing in absolute eternity independent of space and time.

The Universe in which we now live was originally an ideal point, all of whose attributes were equal to ideal zero. It contained absolutely nothing physical, not even pure and weightless energy. In this zero point, which we call the first white space hole, the ideal laws of future nature were embedded. The comprehensive compilation of all these laws constituted a brilliant program for the birth and colossal evolutionary development of the Universe in the future. This ideal program of material development could not have arisen spontaneously, by itself. It could only have been created by an ideal (nonmaterial) and absolutely perfect Creator, whom we call God.

Thus, God split an absolute and ideal zero into a zero sum of material opposites, namely positive and negative energy. God separated the positive energy from the negative energy so that for many billions of years they could not come together and annihilate each other ("And God separated the light from the darkness").

And God was satisfied with the results of this creation, because it contained weightless and formless photons which would be subsequently transformed into weighty substance, which has volume and rest mass ("And God saw the light, that it was good").

The positive energy of the dispersing photons is called the "light," while the negative energy of expanding space is called the "darkness."

And the expanding space arose ("and there was evening") and energetic photons arose ("and there was morning"): thus went the first of the creation of the Universe by the absolutely perfect God. We call this period of the creation the stage of energy evolution: "first day."

## The Second Day

And God created an ideal program of hydrogen evolution. In accordance with this program and in complete harmony with all the laws which constituted this program, the positive energy of photons was transformed into clouds of hydrogen plasma. The compression of these substantial clouds freed up vast expanses of vacuum space, which separated each cloud of hydrogen plasma from all the others. ("And God said, Let there be a firmament in the midst of the waters, and let it divide the waters from the waters.")

Thus, the positive energy of the white holes was transformed into clouds of hydrogen or antihydrogen plasma, which were separated from one another by pure space both within and outside each galaxy. ("And God made the firmament, and divided the waters which were under the firmament from the waters which were above the firmament: and it was so.")

The vast expanses of pure (vacuum) space which separates the clouds of hydrogen plasma are what we call the heavens. And a pure field of vacuum space arose ("and there was evening") and clouds of hydrogen plasma arose ("and there was morning"): thus went the second stage of the creation of the Universe by the absolutely perfect God. We call this period of creation the stage of hydrogen evolution: "second day."

## The Third Day

In accordance with the ideal program of material development of the Universe created by God, one cloud of hydrogen plasma in our Galaxy was compressed so that denser "pieces" separated from it and formed the planets. ("And God said: Let the waters under the heaven be gathered together unto one place, and let the dry land appear: and it was so.") One such planet was our spherical earth, which had depressions where water subsequently flowed, forming the seas and the oceans. ("And God called the dry land Earth, and the gathering together of the waters God called Seas.") The conditions on the primordial earth were sufficiently auspicious for the emergence and development of life ("and God saw that it was good").

And God designed an ideal program of biological evolution. The elements of this program came from another world into the white holes of our Universe. In the white holes, the instructions of this ideal program were converted into material codes at the photon level, and in the clouds of hydrogen plasma they were converted into codes at the level of weighty elementary particles.

And subsequently, the earth produced different species of plants from the weightless and invisible "seeds of life," which it had inherited from the cosmos. After that each species of plant yielded weighty and visible seeds. Each plant was born from a weighty and visible seed, developed, matured, yielded seed after its own kind, aged, died, and was reborn from a seed. ("And the earth brought forth grass, and herb yielding seed after its kind, and the tree yielding fruit, whose seed was in itself, after its kind.")

Thus was laid the foundation for subsequent biological evolution on the earth. ("And God saw that it was good.")

And the earth formed ("and there was evening") and plants appeared on it ("and there was morning"): thus went the third stage of the creation of the Universe by the absolutely perfect God. We call this period of the creation of the Universe the stage of planetary evolution: "third day."

## The Fourth Day

In accordance with the program created by God, the clouds of hydrogen plasma were transformed into stars, the stars combined to form galaxies, and the galaxies formed the Universe. One of those stars was our sun ("the greater light"), which illuminates the earth only in the day time. In our solar system there also arose the moon ("the lesser light"), which together with the stars illuminates the earth only at night. "And God made two great lights: the greater light to rule the day, and the lesser light to rule the night, and he made the stars also. And God set them in the firmament of the heaven to give light upon the earth."

And God was satisfied that the relative configuration and motion of the earth, the sun, and the stars contained all the necessary conditions for biological evolution. "And God saw that it was good."

And the stars began twinkling ("and there was evening,") and the sun began shining ("and there was morning"): thus went the fourth stage of the creation of the Universe by the absolutely perfect God. We call this period of the creation of the Universe the stage of stellar evolution the "fourth day."

## The Fifth Day

And God created an ideal program of biological evolution, by which life would inevitably and logically emerge and develop, at first in the water and then on land. The elements of this program created by God in the Ideal World would be converted in the white holes of the Material World into energetic codes. We call these energetic codes "the seeds of life." In the clouds of hydrogen plasma, these energetic codes would be converted into substantial codes at the nuclear or electron level. Under the auspicious conditions of the primordial earth, the elementary codes would be converted into codes at the molecular level, accompanied by the formation of proteins and nucleic acids.

And God first created living creatures in the water, followed by the large fish and sea animals, and then God created the winged birds which fly through the air. Each species of biological system develops in an evolutionary way from a lower to a higher form according to its own program, which differs from the programs of all the other species and was encoded at the energetic or elementary levels. Originally each living species arose from its own weightless and invisible "seed of life," which differed from the seeds of all the other species. Later on all living creatures were born, developed, matured, yielded offspring after their own kind, aged, died, and were reborn. The Ideal God created living creatures not with material hands or a magic wand, but with "the Word of God," by means of an ideal program and genetic codes, which functioned as natural cybernetic systems. "And God created great whales, and every living creature that moveth, which the waters brought forth abundantly, after their kind, and every winged fowl after its kind."

And God was satisfied with the results of this creation, because the biological systems contained all the necessary conditions for the next stage of the evolutionary development of the Universe, namely the stage of intelligent development. We call this stage the "fifth day."

**The Sixth Day**

And God created an ideal program of biological evolution, according to which living creatures would appear not only in the water but also on dry land. And it was so. Different species appeared as a result of the evolutionary development of the "seeds of life," which contained the material codes of the ideal program created by God at the nuclear or even at the photon levels. From the beginning, each species of living creatures was formed from its own weightless and invisible "seed of life," which differed from the seeds of any other species. Subsequently each species of living creatures multiplied after its own kind by sexual means.

Interspecies breeding (for example, breeding a wolf with a horse, or a human with an ape) produces absolutely no results. Each individual living creature is born, develops, matures, produces offspring after its kind, ages, dies, and is born again.

And God created an ideal program of intelligent evolution, by which a human being, the biological image of the ideal God, would logically and inevitably appear. In contrast to the ideal God, this being must possess a material body, because its purpose is to help God in God's creative activity on the earth. Hence the human species (like God) would have to possess high intelligence. And this high intelligence would give it power over all other species.

And a human being appeared on the earth as the inseparable unity of sexual opposites (a man and a woman), as the material image of a nonmaterial God, as the relative likeness of the Absolute Creator. God's absolutely perfect intelligence is the limit toward which human intelligence will always strive in the process of evolutionary development, but will never reach. The Absolute God created man and woman not with material hands or with a magic wand, but by means of the "Word of God," by means of the genetic codes and natural laws which constitute the ideal program of intelligent evolution. "So God created a person in God's own image, in the image of God were they created; God created them male and female."

In accordance with the ideal program of biological evolution created by God, the different species of plants, seed-bearing herbs, and tree-bearing fruits would multiply over the entire earth from year to year. These plants, herbs, vegetables, and fruits became plant food for the human species. "And God said, Behold, I have given you every herb bearing seed, which is upon the face of all the earth, and every tree, in which is the fruit of a tree yielding seed; to you it shall be for to eat."

Other living species would also live off of plant food, including all the animals of the earth, all the birds of the sky, and all the creeping animals. And it was so. "And to every beast of the earth, and to every fowl of the air, and to every thing that creeps upon the earth, wherein there is life, I have given every green herb for meat: and it was so."

And different animal species arose ("and there was evening"), and intelligent human beings arose ("and there was morning"): thus went the sixth stage of the creation of the Universe by the absolutely perfect God. We call this period of the creation of the Universe the stage of intelligent evolution "the sixth day." And God was satisfied with the results of creation, because in the person of man God had made a worthy helper capable (like God) of creating purposefully and consciously. We call this stage of creation the "sixth day."

We have now looked at a purely scientific model of the creation and evolutionary development of the Universe. But how does it differ from the Biblical model? Analysis and comparison will convince us that the scientific and the Biblical models are the same in terms of their essence and semantic contents. The only difference lies in the form and manner of presentation and in the distinctive features of the ancient and modern languages. The scientific model supports rather than disproves the Biblical model.

The Bible describes the process of the creation of the Universe by the Absolute God in the ancient Hebrew vernacular. It does not provide, however, any scientific explanations. Why? Because the people of the time when the Bible was written were not prepared to receive or able to comprehend any sort of scientific explanations in our meaning of the word. At that time people were not interested in why and how a grain of wheat cast on fertile soil turned into a stalk and how grains once again appeared on the stalk. They were simply interested in getting a good harvest.

While an intuitive belief in God or a blind faith in godlessness may have been acceptable in earlier stages of the development of human society, today we are faced with the need for a scientific explanation of belief. Some 3300 years ago Moses gave the world the Bible and revealed God to people, in our time, Albert Einstein paved the way to the Biblical God through modern science. And it was no accident that atheists and materialists branded Einstein an "idealist" for a long time. After all, his theory of relativity constituted the basis for the theory of the expanding Universe, which in turn ground the fundamental atheistic legend of the "eternity and infinity of the Universe" into dust and ashes. Without this foundation, atheism, like a needle without an eye, can jab and jab, but never sew, and can slander and slander, but never convince.

All religious and Biblical truths are scientifically explicable. Atheists were the ones who fabricated the scientific inexplicability of Biblical truths and Biblical miracles in order to make their war against religion in general and the Bible in particular easier. Religious truths are scientifically explicable and all-powerful simply because they are true. But these explanations require science on the extraordinarily high level, toward which modern humanity is moving, not the primitive and antediluvian level at which atheism is stuck. There are no inexplicable phenomena in the Real World. For example, God can easily explain any phenomenon of the Ideal World as well as the Material World. But the world contains many such phenomena which cannot be explained because of the limitations of human intelligence. For example, we can scientifically prove and explain the existence of God as an ideal category with no material attributes. But we cannot completely reveal the essence of God, eternally existing outside of any space and time.

If a deaf person cannot experience all the delights of music, in no way does it imply that music is altogether inaudible. By perfect analogy, if limited human intelligence cannot provide a complete explanation of a particular truth, this in no way implies that this truth is altogether scientifically inexplicable. The absolute truth is that limit of knowledge toward which limited human intelligence will always strive and asymptotically approach but will never reach. But the limited nature of our capabilities does not mean their complete absence. We can achieve relative truth, even though the truth will never be known to us in its absolutely complete and final form. The scientific explicability of particular phenomena in the Material World does not provide any evidence of the primacy of matter. On the contrary, the primacy of God (or the primacy of the objective idea) can be proven scientifically.

The creation of the Material World by the Ideal God can be explained scientifically, and is in no way a fabricated fairy tale. The atheist fable of the "eternal spontaneous development of uncreated matter" is just such a tale, which cannot be explained scientifically or even in real fairy tales. If you listen to the Russian folk tale of Kashchey the Immortal, even small children will laugh ironically at the legend of Kashchey's "immortality." But why do adult and completely intelligent atheists fail to grasp the logic necessary to laugh ironically at the atheist prejudices of immortal matter?

It is the atheist yarn of the "eternal spontaneous development of uncreated matter" which is an unscientific fairy tale, not the religious truth of the creation of the Material World by the Ideal God. In one of his fairy tales, the Russian poet Alexander Pushkin wrote: "There goes the mortar with Baba Yaga/ Walking all by itself."

So what sort of impression do these verses make on us? Even a scientific materialist would laugh at a fantasy written for small children, because a mortar cannot move under its own power. But why do you find it so easy to keep from laughing when atheists try to make you blindly believe in even bigger fairy tales and yarns by claiming that all inanimate matter, not just a mortar in a fairy tale, can move under its own power?

And the atheist fantasies do not stop here. An atheist is a much bigger yarn-spinner than any fabulist. Atheism claims that matter not only moves but develops all by itself and that it not only changes and develops, but does so logically and expediently, i.e., in accordance with the laws of nature and a predetermined program. It has developed so expediently and purposefully that inanimate and mindless matter has inevitably come to life and begun to think. And all of this has transpired all by itself, without any creator of the laws of nature, without any compiler of the program of development, without any will, and without any intelligence.

To top all of this, atheism claims that mindless matter has supposedly created and refined itself without any intelligent creator, that the zero point became unstable all by itself and gave birth to the Universe all by itself, that fish, birds, animals, and apes appeared all by themselves in the Universe, and that later on certain apes metamorphosed into highly intelligent people all by themselves.

The development of science has systematically discredited every atheist and materialist notion of the world. Important steps in this direction were the discoveries of molecules, atoms, electrons, fields, antisubstance, and white and black holes. Initially molecules were considered indivisible. Then, at different stages of its development, science came to consider atoms, electrons, and fields indivisible. And only now have scientists realized that behind the world of the field stands the world of objective ideas, which exist outside of and independently of subjective (human) consciousness. Under the onslaught of proven scientific facts, atheism has been forced to recognize the idealistic theory of the expansion of the Universe which began at a zero point. As a result of this generally recognized scientific theory, the atheistic fabrication of the eternity and infinity of the Universe has collapsed like a house of cards.

In our era it is atheism (not religion) which has become the greatest impediment to the further development of the natural sciences. Atheists and materialists do not want to see the spiritual dimension of life and are obstructing a conscientious scientific quest for the objective truth in the natural sciences by turning them into a branch of the atheistic faith instead of a branch of science. Ultimately they have had to compare human beings to an "advanced cybernetic machine," or call them soulless, "highly organized matter."

Back in 1977, A. I. Oparin, the prominent scientific atheist and Soviet academician, was forced to admit that "mechanistic materialism has proven to be incapable of arriving at a rational solution to the problem of the origins of life."[1] Modern atheism is completely reliant on Marxist materialism, which formally calls itself "dialectical" but is in fact a mechanistic materialism, because it is in glaring conflict with the laws of dialectics, by which matter must have a nonmaterial dialectical opposite, namely the objective idea.

Modern science is capable of proving all the basic premises of the ancient Bible. It is atheism (not the Bible) which is holding to its blind faith in fabricated fairy tales: such as, the self-motion and expedient self-mutation of inanimate matter, the purposeful self-development of mindless nature, the spontaneous transformation of inanimate substance into living creatures, sorcery without a sorcerer whereby a stupid monkey is transformed into an intelligent human being all by itself, creation without a creator in which the Universe was born and developed all by itself from a zero point, a program of evolution without a programmer, laws of nature without a lawmaker, and so forth. As you can see, it is not so easy to believe in atheist miracles if you think about it.

Here we have cited the most advanced and reliable scientific views of the evolutionary development of the Universe. But the scientific model given here is not the only one even today. There are many other hypotheses, and each of them contradicts all the others. Figuratively speaking, there are as many scientific models of the Universe as there are astronomers in the world. Over the last century hundreds of mutually contradictory scientific theories and hypotheses have come and gone. As Vitaliy Rydnik puts it, "There never have been, never are, and never will be any infinitely powerful theories." Any scientific theory, like any human being, is born, develops, exists for a while, ages, and dies. But the ancient Bible has never changed a single letter and stands like an indestructible rock over which the rivers flow. *Scientific theories come and go, but the Bible remains.* But nevertheless something has changed recently. But what?

The distinctive feature of our modern era is the rapid development of science and technology. Because of this, science has matured to the extent that it is capable of understanding those Biblical truths which it once considered "miracles." Hence the time has come when religion and science can talk to each other in the same common scientific language. If it took science more than 3300 years to arrive at the elementary Biblical truth that light existed before the stars and the sun, then all we can do is be amazed at the brilliance of the author of the Bible, who beat science by 3300 years. The Bible was written back in the age of the bow and arrow. Modern science builds spaceships. An ancient arrow which can overtake a modern spaceship is truly magnificent.

God created the Universe in six evolutionary stages, or six "Biblical days." One Biblical day was on the average equal to two billion terrestrial years. On the seventh day God rested and enjoyed the fruits of creation. "And on the seventh day God ended the work which God had made; and God rested on the seventh day from all the work which God had made."

*     *     *

*The results of our objective comparison and dispassionate analysis of the Biblical model of the creation of the world and the scientific model of the evolutionary development of the Universe have convinced us that modern science corroborates rather than discredits the Bible. Advanced science contradicts atheism, not the Bible. Indeed, advanced science discredits the distorted, antiscientific interpretations of atheism.*

And two stars lit up in the sky far away from each other. One was in the South, and the other in the North; one was female, and the other male. And they rushed toward each other, came together and flared up with a brilliant light, and then they vanished. And were you not, Adams and Eves, men and women, born for each other in the same manner, that you should meet, merge into one, flash with the flame of eternal love, and then leave this physical world?

According to Rabbi Simeon Ben Yohai,
Second century (believed to be the author
of the Zohar, the main work of Kabbalah, or
Jewish Mysticism)

# Appendices

## Appendix A:

## The Seventh Day and the Messianic Age

A s a token of solidarity with God, the Bible offers society a seven-day week. Created in the image of God, human beings must work six days a week and rest on the seventh day. "And God blessed the seventh day, and sanctified it: because on that day God had rested from all his work which God created and made."[1] If we follow the logic of the Bible further, after its sixth millennium of historically conscious development humankind should rest and enjoy itself, like God. The seventh millennium should be an era of universal happiness for all mankind.

According to the Hebrew calendar, we have 240 years left until this day. According to the Christian calendar, it will happen approximately in the year 2240. Then the messiah should come and establish a universal paradise on the earth.[2] What is most interesting is that you and I and everyone else will have the opportunity to live in this splendid time. This will be the subject of a separate book entitled *Body and Soul*

\* \* \*

And humankind extends its trembling hands to God and says: "O God Look at how much evil and violence there are in this world! How much innocent blood has been shed on this mortal earth! How many criminals have gone unpunished! How many trainloads of innocent victims have been sent to meet a cruel death! Oh God, pardon the sinners that we are. But, dear God, why do you permit all of this to happen? Why don't you send the messiah to this poor earth to bring happiness to all people? We want the messiah to come now!"

And the messiah responds by saying: "Do you want me right now? I am not coming to you now because you are still not ready to receive me. You want happiness? It is not only up to God. You must do something about it yourselves. Happiness requires human effort. It requires knowledge. Therefore, I will come to you only when your minds grasp the truth of the essence of your own being. I will only come when you learn how to clearly tell good from evil and victim from perpetrator. When you learn to reward the kindness of the human soul, and punish actual, not imaginary, evil. When your hearts overflow with noble love and no room is left for fierce hatred, even toward those you have to punish, in the name of justice, for the sins and crimes they have committed.

"I will not come to you so long as you seek the company of the cruel and the inhuman so that you may become powerful and rich. I will not come to you so long as you worship false gods, such as a Hitler or a Stalin. I will only come to you when you turn your face to the God who created you, when you incessantly reach out to the helpless and the unfortunate to help them and share their fate with them. I will come to you when your desires will be focused on creative work rather than destruction and annihilation.

"I will come to you when you become the material image of the immaterial God, the relative likeness of the Absolute Creator. As the supreme reward for the perfection you achieve, I will build for you a kingdom of bliss and felicity, justice and harmony, truth and peace, love and reason."

# Appendix B:

## Facts from Astrology

According to astrology and the Kabbalah,[1] various days of the week influence the success of our practical activities in various ways. Of all the stages of the evolutionary development of the Universe, the most important is considered to be the third, when the earth, the dwelling place of mankind, was created. Tuesday corresponds to this stage. For this reason, Tuesday is considered to be the most favorable day of the week.

The next in order of importance is considered to be the stage of the evolution of energy, the first "day" of the creation of the world, when God created and separated the light from the darkness. Sunday corresponds to this stage. For this reason, Sunday is also considered to be a favorable day of the week, but to a lesser extent than Tuesday.

In the second and fourth stages of the evolutionary development of the Universe, positive energy (the purest and clearest light) was transformed into the less perfect form of matter: physical clouds of hydrogen plasma, the stars, and the sun. Monday and Wednesday correspond to these stages. For this reason Monday and Wednesday, are considered to be the most unfavorable days of the week. The primordial clouds of hydrogen plasma were colder and more primitive than the stars and the sun. For this reason, Monday is considered to be worse than Wednesday.

Thursday corresponds to the stage of biological evolution: the fifth day of the creation of the world by God. At this stage of the evolutionary development of the Universe, inorganic (more primitive) matter was transformed into living (a more perfect) form of matter; from the primitive "dust of the earth" living creatures, such as fish and birds, were formed. For this reason, Thursday is considered to be as favorable day of the week as Sunday.

At the sixth stage of the evolutionary development of the Universe, the basic purpose of creation was achieved. God created a human couple, whose relative intellect is like the absolute intellect of God. Friday corresponds to this stage. For this reason, Friday is considered to be as favorable a day of the week as Sunday.

From the point of view of astrology, the degree to which a certain day of the week is favorable is determined by the position of the moon on its circumterrestrial orbit. However, this in no way signifies that our success or failure is completely determined by a certain day of the week, since not only the position of the moon influences earthly conditions, but also that of the sun, stars, and other planets. For this reason, not only the days of the week can influence the success of our practical activities, but a whole extremely complex assortment of astrological and nonastrological factors. An examination of these factors would go far beyond the limits of this book. For this reason, we intend in the future to devote a separate book to these questions.

There is every reason to believe that the numerals 6 and 7 relate to magic numbers. For example, Nazi Germany disappeared from this world in 1945, that is, in the seventh year after *Kristallnacht,* "the night of broken glass," which was directed toward the extermination of the Jews.

# Appendix C:

## The International Scientific Center

The messiah will come and the day of universal happiness will arrive only when the overwhelming majority of humankind learns to understand and be aware of the pain of others and when no one wishes to do to others what he or she would not wish done to oneself. But in order to understand someone else's suffering, one must know the truth, and one cannot know the truth without a high level of intelligence.

Therefore, the day of universal happiness will arrive only when humankind reaches the highest stage of its intellectual perfection. And if you want to bring this day a little closer, please learn the truth from reliable scientific sources and spread the word among your friends and acquaintances. One cannot be happy until one understands one's own soul. One can only understand one's soul and purpose in life when one knows the nature of one's existence. Hence a conscientious quest for the objective truth is no trivial matter but rather the prime necessity and sacred duty of anyone who wishes to achieve true happiness.

Our nonprofit research organization, namely, The International Scientific Center, has set itself the task of conducting a conscientious quest for the objective truth, regardless of the subjective whims of any particular individual, of any particular trends, or of any particular politics. Hence we are asking you to provide our organization with assistance if you believe we are moving in the right direction. By providing aid, you will not only enable our organization to conduct objective and independent research, you will make it possible for us to publish our results. We will assume no other obligations than these

We would like to hear from you on our work in general and on this book in particular. Please, try to formulate all of the positive and negative aspects as briefly and clearly as possible. Please, point out any typos, errors, contradiction, shortcomings and merits you may have noticed.

And, which is most important, please tell us whether this book has influenced your scientific and religious outlook in any way.

The author will be happy to receive any critical or noncritical remarks that facilitate the scientific quest for objective truth.

If anything mysterious has happened to you or you have witnessed phenomena confirming some of the religious stipulations, please tell us about this as briefly as possible, while at the same time as completely and clearly as possible. We also urge you to send in a formal review of this book. It will give us your moral support. Don't forget to put your signature and indicate your full name, address, and any academic degrees or titles you may have. We are also in need of business people who could afford to provide a grant for our books or organize presentations to the public, radio and TV.

Moreover, we are interested in reaching out to publishing houses and retail businesses that can publish and distribute our publications under contract, in other countries, beyond the USA.

If you are a professional artist, we invite you to illustrate our publications. Our books may assure the best publicity for your art. If successful, they will help you to immortalize your name.

The editorial board reserves the right to turn down or publish all or part of the contents of your letters and manuscripts, without any commitments. If you claim copyright or fees, please do not mail your manuscript without first obtaining our consent in writing.

Your reviews and suggestions can be mailed to:

International Scientific Center, CE
Post Office Box 350 567
Brooklyn, New York 11235-0567

# Appendix D:

# Scientific Reviews of the Russian Edition of This Book

## Review of "Creation and Evolution" by Joseph Davydov

"The eternal self-development of matter, which was created by no one" is the cornerstone of the atheistic attempt to explain the nature of everything in existence. In the period during which atheism was the reigning ideology of a whole array of so-called Marxist governments, this tenet was not proven and was not substantiated, but was essentially accepted as an axiom for all of the natural sciences. Religious substantiation of creation was also, without proof, declared to be fallacious, fictitious, and not deserving of attention, nor well-grounded on criticism. Such in particular was the attitude towards Biblical explanation of World Creation.

The work of Joseph Davydov is devoted to the debunking of this tenet. At the foundation of his conception of the world lies the idea that in absolute eternity, outside of time and of any space there exists not matter, but an Absolute God. He specifically is primary in relation to all of the Relative World, not only the physical but also the ideal. And He specifically, the Absolute God, was the only real creator of the universe, having completed His creation in the six days of the Bible, bearing in mind that one Biblical day is equal on the average to two billion terrestrial years.

In accordance with these views, on the "first day," "the evolution of energy" took place; on the "second"—"hydrogen evolution;" on the "third"—"planetary evolution ;" on the "fourth"—"stellar evolution;" on the "fifth"—"biological evolution;" and on the "sixth day"—"intellectual evolution."

Let us dwell on the last two "days." The ideal program of biological evolution created by God was supposed to be inevitably completed, and actually was completed through the origin and development of life, first in water and later on land. Does this idea contradict the data of present-day natural sciences? Joseph Davydov maintains that it does not contradict them, and he proves his point quite convincingly. The present-day level of molecular-biological knowledge leads us to believe that life originates during the merging of three "flows"—"the flow of matter" (protein), "the flow of energy" (the processes of reduction-oxidation), and "the flow of information" (the genetic code of the nucleic acids). The most complex in terms of its comprehension is the origin of the information flow.

Advanced biology is not able to explain the primary origin of the genetic code—its original "intelligent contents" and its capacity for efficient evolutionary complexity. And in this connection the author's statements—that the elements of the program of the evolutionary complexity of life on earth were created by the Absolute God in an Ideal World, and were afterwards reshaped in the "white holes" of the physical world into energy codes within the depths of clouds of hydrogen plasma, which were later transformed into "substance" codes at the nuclear and electronic level—seem very arguable. The real means of the material evolution of nucleic acids and proteins, their influence on each other and their formation of cells is and will be an object of thorough research in all succeeding times, but it is impossible not to agree with the thought that all of this research will be directed towards the aim of more clearly and distinctly understanding the elements of that primordial program of the Creator which was laid down by Him and was realized on "the fifth day of creation." Atheistic natural science cannot cope with this task.

The natural continuation of the "biological evolution of the fifth day" is the "intellectual" (in Joseph Davydov's terminology) evolution of the "sixth day." The appearance of various species of living creatures is regarded by the author as a result of the evolutionary development of "the seeds of life" which contain at a nuclear or photon level the material codes of the ideal program created by God. One cannot fail to notice how in the text of the Bible the expression "according to his kind," as applied to every category of living creatures, is constantly repeated. In this sense, the Bible is an authentic guide to the general genetics of life on earth. How much more profound is such a description of life than the "neo-Lamarckian" mechanistic ideas about mutability, which reached the level of pagan superstition in the so-called "Lysenko biology," which comparatively recently reigned in the former USSR.

The program of intellectual evolution naturally and inevitably led to the appearance of man - the biological image of the Ideal God, with His freedom of choice and ability to create. And again these ideas to a considerably greater degree correspond to the total knowledge of present-day science than do primitive, pseudo-materialistic pieces of information about the linear evolution of man from man-like primates.

On the whole it is impossible not to agree with the author that the results of objective analysis confirm, rather than deny, the Biblical history of the creation of the world, and particularly of the biological forms of life on earth. The work of Joseph Davydov in our opinion substantially helps the reader to free himself from the ways of vulgar atheism and to come closer to God!

Daniel Golubev,
Doctor of Medical Sciences
Professor of Virology
Full Member of American Society for Virology,
Bio-Virus Research Society for Virology,
Bio-Virus Research Incorporated,
Vice President Research and Development,
International Association of Art and Science,
Executive Vice President.

## Review of the work of Joseph Davydov "The Laws of the Preservation and Creation of Matter"

From the point of view of contemporary physics, two processes which are opposite in direction are taking place in nature: the disappearance and beginning in pairs of elementary particles and antiparticles such as electrons and positrons, neutrinos and antineutrinos, protons and antiprotons, etc. Examining a considerable number of analogous examples and using the generally recognized method of inductive cognition of objective truth, Joseph Davydov in his work has reformulated two fundamental laws of nature: the conservation of matter and the creation of matter, which are not in conflict but rather complement and substantiate each other. The merit of Joseph Davydov lies in the fact that he has succeeded in clearly formulating the possibility of the creation of matter in the form of a zero sum of non-zero opposites. On the whole, we agree with the author's original conclusion that God split zero into a zero sum of positive and negative energy—into "light" and "darkness" which were divided one from another. The author's point of view is close to the position of Spinoza, Newton, Einstein, and Bohr.

> Genrikh Golin,  Professor of Physics,
> Doctor of Sciences,
> President of the International Association of Activists in Science and Culture.

## Review of Joseph Davydov's "Creation and Evolution"

The problem of the interrelationship of science and religion remains an acute and topical one, not only in terms of the dispute between atheism and religion, but also for science as such.  Evidence  for this is the numerous publications and conferences (one of the latest, under the heading of "Critical Dialogues in Cosmology" took place recently in Princeton, New Jersey) oriented towards specialists, and also frequent publications in the mass media (The New York Times, "Yevreyskiy Mir," and others), oriented to the general reader. The difference between an article published in a newspaper and the systematic exposition of material in a detailed monograph is quite evident. For this reason the attempt by the author to give a detailed analysis of the Biblical model of the creation of the world and to compare it with the contemporary scientific picture of the world is timely. In my opinion, the author has succeeded in expounding his point of view on quite a high scientific level, and in precise and clear literary Russian.

It is to be hoped that the monograph under review will be well received by readers, especially that part of them who through enforced loyalty to the reigning ideology were limited in their access to materials of this kind.

Guennadi Koulechov
Professor of Physics, Doctor of Sciences

**Review of the Chapter "The Fifth Stage of the Creation of the World: Biological Evolution."**

At the present time the natural sciences hold to the evolutionary model of the origin and development of life on earth, although this problem still needs to be further resolved. The author Joseph Davydov unequivocally persuades the reader that the scientific theory of biological evolution confirms, rather than negates the Bible. Against the Bible stands forth not the scientific theory itself, but its antiscientific atheistic interpretation, which maintains that the efficient evolution of a living organism takes place, as if it were, by itself, spontaneously, without any program and without an intellectual creator. The processes of the information-energy transformation of inorganic matter into living organisms which are set forth in the book cannot fail to attract the serious reader through the boldness of the author's views and his original approach to the problems he examines.

On the whole, the book is intended for a fairly well-educated reader, although the graphic quality of its presentation and its enthusiasm make the text more understandable for everyone who is interested in the subject of the origin of life and intelligence on earth.

Tamara Vyshkina
Candidate of Chemical Sciences

# Notes

## Chapter 1: The Language of the Bible and Science

1. A Dictionary of Philosophy (Moscow: Politizat, 1975), 266.
2. See, for example, A. I. Oparin, *Matter, Life, Intelligence* (Moscow: Nauka, 1977).
3. Albert C. Bauch, *A History of the English Language* (New Jersey, 1993), 401-403.
4. Genesis 1:7. All Biblical quotations are based on the King James version, with archaic language modernized by the editor.
5. Cyril Ponnamperuma, *The Origins of Life* (Moscow: Mir, 1977), 12.
6. In *The Science of Secrets*, Vol. no. 3, 123.
7. *Dictionary of Philosophy*, 29.
8. Ibid.

## Chapter 2: Substance and Science

1. *Translator's note*: The Russian word *veshchestvo* has a variety of English translations, including "matter," "substance," "material," and so forth, and generally refers to "physical matter." The author of this book, however, proposes to use the English term "substance" to refer exclusively to a physical reality whose rest mass is not equal to ideal zero, even though it may be infinitesimally small or infinitely large. In cotrast to "substance," the author proposes to use the term "matter" to refer to any physical reality (weighty or weightless), even if its rest mass is equal to ideal zero.
2. N. L. Glinka, *General Chemistry, A Textbook for Soviet Institutions of Higher Education* (Moscow, 1978), 20.
3. V. I. Rydnik, *The Laws of the Atomic World* (Moscow: Atomizdat, 1975), 332-340).
4. *General Chemistry*.
5. In the Russian original, the word materiya is used. Refer to note 1 for explanation.
6. Albert Einstein, *Collected Scientific Works*, Volume 1 (Moscow: Nauka,

1965), 679.
7.  Joseph Soulson, *Worlds* (New York: International Scientific Center, 1991), 54.
8.  See M. A. Markov, *On the Nature of Matter* (Moscow 1976), or Joseph Shklovsky, *The Universe, Life, and Intelligence* (Moscow: Nauka, 1975), 205-206.

### Chapter 3: Particles and Antiparticles

1.  *Laws of Atomic World*, 258-260.
2.  A. P. Trofimenko, *The Universe: Creation or Development?* (Minsk, 1987), 138.

### Chapter 4: The Fumdamental law of Nature

1.  *Control, Information, and Intelligence* (Moscow: Mysl, 1976), 232.
2.  *Worlds*, 44.
3.  See Henri Poincaré, *The New Mechanics* (St. Petersburg, 1911), 41-42.
4.  See *Dialectical and Historical Materialism* (Moscow: Politizdat, 1974), 81-82.
5.  *Control, Information, Intelligence*, 232.
6.  *The Universe: Creation or Development*, 138.

### Chapter 5: The Laws of Conservation and Creatablility of Matter

1.  *Laws of the Atomic World*, 297.
2.  Ibid, 286, 296.
3.  Ibid, 286, 296.
4.  Ibid, 297, 298.
5.  Ibid, 286, 296.
6.  *Control, Information, and Intelligence*, 44.
7.  *Laws of the Atomic World*, 259.

## Chapter 6: Idea and Mattter

1. *The Universe*, 138.
2. M. A. Markov, *On the Nature of Matter* (Moscow, 1976); *Worlds; The Universe.*
3. *The Universe*, 311.
4. *Worlds*, 323.
5. See *Worlds*.

## Chapter 7: The Theory of the Expansion of the Universe

1. To learn more about this loathsome document of the twentieth century, which is part and parcel of the history of "scientific atheism," see the *Great Soviet Encyclopedia*, 1953, Vol. 23, 112.
2. S. T. Melyukhin, *The Problem of the Finite and the Infinite* (Moscow: Gospolitizdat, 1958), 95.
3. *The Fundamentals of Marxist Philosophy: A Textbook* (Moscow: Politizdat, 1962), 95.
4. A. I. Kitaygorodsky, *Physics for Everyone*. Photons and Nuclei (Moscow: Nauka, 1979), 190-193.
5. *The Universe*, 92.
6. B. Bova, *The New Astronomy* (Moscow: Mir, 1976), 217; or: *The Universe*, 60, 68-70.
   Also: Indian graduate student Subrahmanyan Chandrasekhar calculated that a cold star of more than about one and a half times the mass of the sun would not be able to support itself against its own gravity. A similar discovery was made about the same time by the Russian scientist Lev Davidovich Landau.
7. *Worlds*, 221-230.
8. Introduction to Philosophy. A Textbook for Institutions of Higher Education in the USSR (Moscow: Politizdat, 1990), Part 2, 61; Physics for Everyone, 21, 38, 46, 47.

## Chapter 8: The Absolute God and the Relative World

1. Genesis 1:1.
2. Genesis: 1:8.
3. Yemelyan Yaroslavsky, *The Bible for Believers and Non-Believers* (Moscow: Gospolitizdat, 1958).
4. *Worlds*, 363.
5. Ibid, 365.
6. *Worlds*.
7. Ibid, 265-267.
8. *Worlds*.
9. Ibid.
10. Ibid.
11. *The Universe: Creation or Development*, 138.

## Chapter 9: The Theory of the Evolutionary Universe

1. *Fundamentals of Marxist Philosophy*, 193.
2. Ibid, 190.
3. *Worlds*, 265-285.
4. *The Universe*, 93.
5. Ibid, 92.
6. See pages 62-80.
7. *Worlds*, 227, 267, 277, 368.
8. *Science of Secrets*, Vol. 4, 73.
9. *Worlds*, 309.

## Chapter 10: The First Stage of Creation

1. Genesis 1:1-5.
2. *Dialectical Materialism: A Textbook* (Moscow: Mysl, 1989), 96; Yu. N. Yefremov, *In the Depths of the Universe* (Moscow: Nauka. 1977), 61; *Physics for Everyone*, 93; I. D. Novikov, *The Evolution of the Universe* (Moscow: Nauka, 1979), 96, 158; *The Universe*, 68, 91.
3. See page 103.

4.  E. P. Levitan, *The Physics of the Universe* ((Moscow: Nauka, 1976), 151.
5.  See, for example, *The New Astronomy; Physics for Everyone; Evolution of the Universe; Worlds; The Universe;* Joseph Shklovsky, *The Stars: Birth, Life, and Death* (Moscow: Nauka, 1975).
6.  *Worlds,* 259.
7.  *The Stars,* 14.
8.  See *The Bible for Believers and Non-Believers,* 19.
9.  *Evolution of the Universe,* 113.
10. *The Stars,* 14.

## Chapter 11: The Second Stage of Creation

1.  Genesis 1:6.
2.  Genesis 1:1-8.
3.  *Evolution of the Universe,* 113-114.
4.  Genesis 1:6.
5.  Genesis 1:7.
6.  *Fundamental of Marxist Philosophy,* 88.
7.  See page 46.
8.  *Laws of the Atomic World,* 294.
9.  Einstein's *Collected Scientific Works,* 679.
10. Genesis 1:6.

## Chapter 12: The Third Stage of Creation

1.  Genesis 1:9-13.
2.  See *The Universe,* 117-134.
3.  Ibid, 118-120.
4.  *The New Astronomy,* 187.
5.  V. A. Ugarov, *Special Relativity Theory* (Moscow: Nauka, 1977), 134.
6.  *The Universe,* 133; and *Secrets of Science,* Vol. 4, 73-74.
7.  *The Universe,* 155.
8.  Ibid.
9.  Genesis 1:9.
10. Genesis 1:10.

11. Genesis 1:11-12.
12. See *The Bible for Believers and Non-Believers*, 31.
13. *The Universe*, 199.
14. *Matter, Life, Intelligence*, 31.
15. *The Bible for Believers and Non-Believers*.
16. Khimiya i zhizn [Chemistry and Life], 1973, No. 1.
17. Genesis 1:9-10.
18. Genesis 1.

## Chapter 13: The Fourth Stage of Creation

1. Genesis 1:14-19.
2. *The Universe*, 81.
3. See *Worlds*, 309-331.
4. *Physics of the Universe*, 45.
5. Ibid, 48.
6. See Sergey Ivanov's article in the periodical *Panorama*, No. 695, 29.
7. Genesis 1:14.
8. Genesis 1:15.
9. Genesis 1:16.
10. See Einstein's *Collected Scientific Works*, 535.
11. Genesis 1:17.
12. Genesis 1:17-18.
13. *The Bible for Believers and Non-Believers*, 21-22.
14. Genesis 1:14-16.
15. *The Universe*, 97.

## Chapter 14: The Fifth Stage of Creation

1. Genesis 1:20-23.
2. *The Universe*, 146.
3. *The Universe*, 163; *Science of Secrets*, Vol. 5, 136-139.
4. *Matter, Life, Intelligence*, 122
5. R. Glazer, *Biology in a New Light* (Moscow: Mir, 1978), 164.
6. *Matter, Life, Intelligence*.

7.  See article by Saveliy Kamenitskiy in *Novoye Russkoye Slovo*, January 26, 1996.
8.  See V. Pekelis, *The Cybernetic Blend* (Moscow: Znaniye, 1973), 65.
9.  Ibid, 67.
10. *Matter, Life, Intelligence*, 198.
11. Genesis 1:20.
12. Genesis 1:21.
13. Ibid.
14. Genesis 1:22.
15. *Matter, Life, Intelligence*, 22-23.
16. The Universe, 163.

## Chapter 15: The Sixth Stage of Creation

1.  Genesis 1:24-31.
2.  Genesis 2:21
3.  Genesis 2:23.
4.  A. Krayev, *Human Anatomy* (Moscow: Meditsina, 1978), Vol. 2, 202.
5.  See, for example, *New Astronomy*, 34.
6.  *Cybernetic Blend*, 59.
7.  Genesis 1:24.
8.  Genesis 1:25.
9.  Genesis 1:26.
10. Genesis 1:27.
11. Genesis 1:28.
12. Genesis 1:29.
13. Genesis 1:30.
14. *The Bible for Believers and Non-Believers*, 36.
15. Genesis 1:26.

## Conclusion

1.  *Nature of Matter*, 25, 198.

## Appendix A

1. Genesis 2:3.
2. *Worlds*, 344.

## Appendix B

1. Philip Berg, *Astrology-Kabbalah* (New York, 1986).

# BIBLIOGRAPHY

## HEBREW

*Bereshit* (Book of Genesis).

## RUSSIAN

Bova B., *The New Astronomy*, Mir, Moscow, 1976.

*Introduction to Philosophy*: A Textbook for Institutions of Higher Education in the USSR. Politizdat, Moscow, 1990, Part 1, 368 pp.

*Introduction to Philosophy*. A Textbook for Soviet Institutions of Higher Education. Politizdat, Moscow, 1990, Part 2, 640 pp.

Vilenchik M. M. *The Biological Foundations of Aging and Longevity*. Znaniye, Moscow, 1976, 160 pp.

Hegel, George, Wilhelm Friedrich. *The Encyclopedia of Philosophy*. Mysl, Moscow, 1977.

Glazer R. *Biology in a New Light*. Mir, Moscow, 1978, 174 pp.

Glinka N. L. *General Chemistry*. A Textbook for Soviet Institutions of Higher Education. Moscow, 1980, 720 pp.

Gorelik G. S. *Oscillations and Waves*. Fizmatgiz, Moscow, 1959, 572 pp.

*Dialectical and Historical Materialism*. Politizdat, Moscow, 1974, 368 pp.

*Dialectical Materialism*: A Textbook for Soviet Institutions of Higher Education. Vysshaya shkola, Moscow, 1987, 336 pp.

*Dialectical Materialism*: A Textbook. Mysl, Moscow, 1989, 400 pp.

Dubinin N. P., *Genetics and Man*. Prosveshcheniye, Moscow, 1978, 144 pp.

Yefremov Yu. N. *In the Depths of the Universe*. Nauka, Moscow, 1977, 224 pp.

Kitaygorodsky A. I. *Physics for Everyone*. Photons and Nuclei. Nauka, Moscow, 1979, 208 pp.

Komarov V. N. *Atheism and the Scientific Picture of the World*. Prosveshcheniye,Moscow, 1979, 192 pp.

Kompaneyets A. S. *What Is Quantum Mechanics?* Nauka, Moscow, 1977, 216 pp.

Kosidovsky Z. Biblical Tales. Politizdat, Mos-cow, 1987, 464 pp.

Krayev A. V. Human Anatomy in Two Volumes. Meditsina, Moscow, 1978.

Levitan E.P. Astronomy, a textbook for high scools. Prosvewenie, Moscow, 1994, 208 pp.

—*The Physics of the Universe.* Nauka, Moscow, 1976, 200 pp.

Luzin N. N. *Differential Calculus.* Nauka, Moscow, 1952, 476 pp.

Markov M. A. *On the Nature of Matter.* Moscow, 1976.

*Materialistic Dialectics.* Politizdat, Moscow, 1985, 352 pp.

Melyukhin S. T. *The Problem of the Finite and the Infinite.* Gospolitizdat, Moscow, 1958, 264 pp.

*Scientific Atheism:* A Textbook for Soviet Institutions of Higher Education. Politizdat, Moscow, 1976, 288 pp.

Novikov I. D. *The Evolution of the Universe.* Nauka, Moscow, 1979, 176 pp.

*The Fundamentals of Marxist Philosophy:* A Textbook. Politizdat, Moscow, 1962, 656 pp.

*The Fundamentals of Marxist-Leninist Philosophy.* A Textbook. Politizdat, Moscow, 1976, 464 pp.

Oparin A. I. *Matter → Life → Intelligence.* Nauka, Moscow, 1977, 208 pp.

Pekelis V. *The Cybernetic Blend.* Znaniye, Moscow, 1973, 240 pp.

Ponnamperuma S. *The Origins of Life.* Mir, Moscow, 1977, 176 pp.

Rydnik V. I. *The Laws of the Atomic World.* Atomizdat, Moscow, 1975, 370 pp.

Smirnov V. I., *A Course in Higher Mathematics.* Volume 1. Fizmat, Moscow, 1965, 480 pp.

—*A Course in Higher Mathematics.* Volume 2. Fizmat, Moscow, 1965, 656 pp.

—*A Course in Higher Mathematics.* Volume 3, Part 1. Fizmat, Moscow, 1965, 324 pp.

—*A Course in Higher Mathematics.* Volume 3, Part 2. Fizmat, Moscow, 1969, 672 pp.

Soulson, Joseph. *Worlds.* International Scientific Center, New York, 1991, 400 pp.

*The Science of Secrets.* Israel, 1982.

Taylor, P. J. *The Origin of the Chemical Elements.* Mir, Moscow, 1975, 230 pp.

Trofimenko A. P., *The Universe: Creation or Development?* Belarus, Minsk, 1987, 160 pp.

Tursunov Akbar. *Philosophy and Modern Cosmology.* Politizdat, Moscow, 1977, 192 pp.

Ugarov V. A. *Special Relativity Theory.* Nauka, Moscow, 1977, 384 pp.

*Control, Information, and Intelligence.* Mysl, Moscow, 1976, 384 pp.

A *Dictionary of Philosophy*. Politizdat, Moscow, 1975, 496 pp.

Shklovsky, Joseph. *The Universe, Life, and Intelligence*. Nauka, Moscow, 1976, 340 pp.

—*The Stars: Birth, Life, and Death*. Nauka, Moscow, 1975, 368 pp.

Shulman, Solomon. *Extraterrestrials Over Russia*. Hermitage, USA, 1985, 208 pp.

Einstein, Albert. *Collected Scientific Works:* Volume 1. Nauka, Moscow, 1965, 700 pp.

Erdei-Gruz T. *Chemical Sources of Energy*. Mir, Moscow, 1974, 304 pp.

—*Fundamentals of the Structure of Matter*. Mir, Moscow, 1976, 488 pp.

Yaroslavsky, Yemelyan. *The Bible for Believers and Non-Believers*. Gospolitizdat, Moscow, 1958, 408 pp.


**ENGLISH**


Albert C. Bauch. *A History of the English Language*, New Jersey, 1993.

Berg Philip. *Astrology-Kabbalah*, New York, 1986, 256 pages.

Hawking, Stephen. *A Brief History of Time from the Big Bang to Black Holes*. Bantam Books, New York, 1988, 198 pp.

Sagan, Carl. *Science as Candle in the dark*. 1996, 460 pages.

—Pale blue dot. Balentine Books, New York, 1994, 362 pages.

*Seventeenth Texas Symposium on Relativistic Astrophysics and Cosmology*. The New York Academy of Sciences, New York, 1995, 728 pp.

# GLOSSARY

**Absolute** - Perfect in nature and quality.

**Absolute God** - the absolutely perfect, single, and intellectual Creator of the entire relative and multifaceted world. God exists in absolute eternity outside of any space and time. Space and time are a relative product of the creative activity of the Absolute God and are not a sphere of God's own existence. For this reason, atheistic questions such as: "Where can God be found?" or "Who created God?" are senseless and devoid of any scientific content.

Being the absolute opposite of relative matter, God does not contain any physical attributes whatsoever. God has no physical body, no hands, feet, eyes, ears, or anything physical at all. Physical bodies are the product of the creative activity of the Absolute God but not God's component elements. For this reason, any kind of atheistic claims such as: "If God exists then show Him to me..." are senseless and devoid of any scientific content.

The Absolute God is the Creator of relative space, relative time, and of everything which moves and changes in space and time.

**Actuality** = Acting reality.

**Annihilability of energy** (уничтожимость энергии) - The possibility to nullify both photon and antiphoton (positive and negative energy) under certain conditions.

**Annihilability of energy** (уничтожимость энергии) - The possibility of simultaneous conversion both photon and antiphoton to an ideal nothing under certain conditions.

**Annihilability of matter** (уничтожимость материи) - The possibility to nullify both positive and negative components of material opposites under certain conditions.

**Annihilation** - The simultaneous disappearance of opposites (positive and negative energy; an electron and a positron; a particle and an antiparticle; a substance and an antisubstance; a buyer and a seller; etc.)

**Annihilation of energy** (уничтожение энер-гии) - The simultaneous conversion both positive and negative energy to ideal nothing under certain conditions.

**Antielectron** - The electrical and energetical opposite of an electron.

**Antihydrogen plasma** (антиводородня плазма) - The electrical opposite of hydrogen plasma, consisting of positrons and electro-antiprotons.

**Antineutron** - The energetical opposite of a neutron possessing a negative rest mass.

**Antiparticle** - An electrical or energetical opposite, or any other kind of physical opposite of an elementary particle (antielectrons, antiprotons, antineutrons, antiphotons, antineutrinos, etc.).

**Antiphoton** - The energetical opposite of a photon. The vacuous space of the Universe consists of an entire continuum of invisible and unweighable antiphotons ("the darkness" according to the Bible).

**Antiproton** - The electrical or energetical opposite of a proton.

**Antisubstance** (антивещество) - The electrical or energetical opposite, or any other opposite, of substance which possesses a rest mass and a weight (positrons, antiprotons, antineutrons, anti-atoms, antimolecules, and everything which can be constructed from them).

**Anti-Universe** - The energetical opposite of the Universe where space is produced from positive energy and weighable matter is produced from negative energy.

**Artificial Intellect** (искусственный интеллект) - The physical opposite of ideal intellect (the cybernetic system which can be created by human).

**Astrology** - The study professing to fortell the future and intepret the influence of the heavenly bodies upon the destinies of men.

**Astronomy** - The science which studies the structure, motion, and development of celestial bodies and systems.

**Astrophysics** - The science that studies the properties of cosmic substances and fields.

**Atheism** - The fallacious opposite of true religion (a blind belief in the fantastic absence of God).

**Atheistic prejudices** (атеистические предрассудки) - The initial premises of atheism, which stand in glaring contradiction to the laws of nature, but which remain in the consciousness of unsophisticated people as a result of their scientific backwardness. For example, a backward person considers everything which does not have volume or weight to be an impossible category. On this basis he or she acquires the delusive certainty that an incorporeal and invisible God supposedly could not exist. However, contemporary science has authenticated the existence of unweighable and invisible physical fields, such as radio-waves for example. The weight, volume, and all the dimensions of a photon, an antiphoton, a neutrino and an antineutrino are equal to zero.

**Atheistic superstition** (атеистическое суеверие) - The naive belief in atheistic fables, which are in glaring contradiction to the laws of nature: the eternal and infinite nature of the Universe, the self-creation and efficient self-development of mindless nature, the spontaneous transformation of inorganic matter into a living creature, etc.

**Atlantis** (Атлантида) - A vast, fertile, densely populated island that existed in bygone times in the Atlantic Ocean west of Gibraltar and which sank to the bottom of the ocean as punishment for the serious sins committed by its inhabitants.

**Atom** - A component of a molecule which represents the smallest model of the immense solar system. The nucleus of the atom consists of protons and neutrons, and electrons revolve around the nucleus as does the Earth around the sun.

**Attribute** - An immutable property of an object. If the attribute is lost, then the object itself ceases to be.

**Biblical Day** - An individual stage in the creation of the Universe by God, which according to present-day calculations is equal to approximately two billion terrestrial years.

**Big bang** (большой взрыв) - The birth of a colossal amount of pure and weightless physical energy from the first cosmic white hole. The Big Bang started suddenly and immediately with the same colossal speed as the speed of light.

**Biological cells** - The "elementary building blocks" from which all biological systems and organisms are built.

**Biological evolution** - The nonrevolutionary process of the gradual formation from inorganic matter ("from the dust of the earth") of various types of plants and living organisms, which differ from each other qualitatively and reproduce within their genus but which cannot crossbreed. According to the Bible: the fifth day (stage) of the creation of the Universe by God.

**Biological life** - The dynamically stable existence of biological cells, systems, and organisms (the unconscious opposite of conscious life). Biological life is not life to the same extent that artificial (physical) intellect is not ideal intellect.

**Biological organism** - An individual and dynamically stable unity of a certain number of biological cells or systems, the division of whose component parts is not possible without the organism's losing its individuality.

**Biological species (биологический вид)** - The category of living creatures whose breeding yields fertile offsprings.

**Biological system** - The efficient and dynamically stable unity of a certain number of biological cells, which have a common purpose and which cannot exist one without the other.

**Bionics (бионика)** - Application of biological principles to the design and study of engineering systems.

**Black cosmic hole** - An ideal point in physical space, in which the annihilation (simultaneous disappearance) of an equal quantity of the positive energy of a dying star or galaxy and the negative energy of surrounding space is fully completed.

**Cell** = A biological cell.

**Code (код)** - The physical record of ideal information.

**Codex (кодекс)** - The comprehensive and systematically arranged collection of laws (in written or coded form).

**Cognizer (субъект)** - A human being involved in the cognitive activity of objective reality (the opposite of object).

**Collapse** - The irreversible loss of stability, which leads a system to inescapable self-destruction (viz. gravitational collapse).

**Competitive opposites (конкурентные проти-воположности)** - The multitude of opposites that struggle for the same place or for the right to exist.

**Component opposites (компонентные проти-воположности)** - The opposites that must exist simultaneously.

**Continuum (сплошная непрерывность)** - The continuous extent in which no part of which can be distinguished from adjacent parts.

**Cosmology** - The science which studies the structure and development of the Universe as a whole.

**Creatability of energy (сотворимость энергии)** - The capability of matter to be created under certain conditions.

**Creatability of matter (сотворимость материи)** - The capability of matter to be created under certain conditions.

**Critical density** - The density of a physical body which is low to the point where gravitational forces become so powerful that not a single physical particle (not even a photon) can break away from its surface. With future growth in density, the critical state of the body is strengthened.

**Critical radius** - The radius of a physical body which is large to the point where gravitational forces become so powerful that not a single physical particle (not even a photon) can break away from its surface. With future reduction of the radius, the critical state of the body is strengthened.

**Critical speed** = The speed of light.

**Darkness** (тьма) - The direct opposite of light, negative opposite of positive energy, negative energy.

**Day** (день) - opposite of night.

**Day** (сутки) = Terrestrial day = Terrestrial *Twenty-Four-Hour Period* (земные сутки) - The interval of time in the course of which the Earth performs one complete rotation around its own axis.

**Deduction** - A method of scientific inference in the course of which from a general law follows a particular result.

**Density** - The quantity of mass enclosed in a unit of volume.

**Destructability of matter** (разрушимость мате-рии) - The capability of matter to be destroyed under certain condiions

**Dialectics** - The science of the most general laws of motion, change, and development of the real world.

**Dialectical opposites** (диалектические про-тивоположности) - The opposites that are bound to be periodic and to follow sequentially, one after the other with the passage of time, such as day and night.

**Dimension** (измерение) - Any of the least number of independent coordinates required to specify a point in space uniquely.

**Earth** (Земля) - The third planet in order of distance from the Sun and fifth in order of size: the physical abode of humankind.

**Earthly** = Terrestrial (земной).

**Electro-antiparticle** (электро-античастица) - The electrical opposite of elementary particle.

**Electro-antisubstance** (электро-антивещество) - The electrical opposite of substance, possessing a positive rest mass (positrons, electro-antiprotons, electro-antiatoms, and everything which is constructed from them).

**Electron** - An elementary particle which possesses a rest mass of $9.11 \times 10^{-31}$ kg and an elementary negative charge of electricity.

**Elementary antiparticles** = Antiparticles

**Elementary particles** - The smallest physical elements which in all known processes behave as a single whole (photons, electrons, protons, neutrons, etc.).

**Energo-antiparticle** (энерго-античастица) - The energetical opposite of elementary particle.

**Energo-antisubstance** (энерго-антивещество) - The energetical opposite of substance, possessing a negative rest mass (antielectrons, antineutrons, energo-antiprotons, energo-antiatoms, energo-antimolecules, and everything which can be constructed from them).

**Energo-microcivilization** - A community of intellectuals which, according to the hypothesis of M. A. Markov, can exist in the depths of unweighable photons and invisible antiphotons.

**Energy** = Physical energy, unless specifically indicated otherwise.

**Energy evolution** - The non-revolutionary process of the natural birth from nothing of an equal quantity of positive and negative energy ("of light and darkness"). According to the Bible: The first day (stage) of the creation of the Universe by God.

**Evolution** - A nonrevolutionary development by stages in which each succeeding stage differs qualitatively from the preceding one (the opposite of revolution).

**Field** - An unweighable ocean of wave energy: an electromagnetic field, a light field, gravitational field, biological field, etc.

**Firmament** - The expanse of the heavens (просторы небесные).

**First white cosmic hole** (первая белая косми-ческая дыра) - The ideal nonphysical point from which our Universe was born and is expanding. The Universe was born from the first cosmic white hole as a zero sum of positive and negative energy (according to the Bible – "light" and "darkness").

**Friedmons** - Elementary particles in which living or intelligent creatures exist.

**Galactic day** (галактичиесий день) - The interval of time during which the galaxy completes one full rotation around its own axis.

**Galactic year** (галактичиесий год) - The interval of time during which the galaxy would complete one full revolution around the center of the Universe if the Universe were to continue to exist for all that time. However, according to the scientific theory of Joseph Davydov, the Universe during the entire time of its existence, completes a revolution around its axis of only 180 degrees, after which it is transformed into the Anti-Universe, similar to day on Earth turning into night.

**Galaxy** - An accumulation of a large quantity of stars, forming a single cosmic system. Our galaxy contains 150 billion stars, one of which is the sun.

**Genesis** (возникновение) - A coming into being.

**Genetic code** - The molecular record of a genetic program which definitively determines the norm of behavior of one or another biological cell or one or another particular organism (a physical category, transmitted by heredity).

**Genetic program** - The meaningful content of a genetic code (an ideal category).

**Genetic Selection** = Natural selection.

**God** = the Absolute God = The Absolute Creator of the relative world.

**Gravitational collapse** - The irreversible loss of stability in the existence of a cosmic system (stars or galaxies) as a result of the gravitational forces of compression exceeding the inert forces of expansion. As a result, the cosmic system is compressed to zero and disappears completely from this world, but it does not transform into any other form.

**Happiness** - A conscious approach to absolute perfection.

**Hatred** - The worst form of mental illness, which inevitably leads to destruction and ruin.

**Human being** - A physical image of the nonphysical God and a relative likeness of the absolute Creator. A human being is obliged to love and create, as does God. In this lies his or her fundamental essence. If a human being hates and destroys then he or she becomes the opposite of God, and not God's likeness. For this reason, such a human being ceases to be a human being.

**Hydrogen** (водород) - The lightest and at the same time the most common chemical element ($H_2$) in nature. A molecule of water consists of two atoms of hydrogen and one atom of oxygen ($H_2O$).

**Hydrogen evolution** - The nonrevolutionary process of the gradual and lawful ("natural") conversion of primordial positive energy into individual clouds of hydrogen plasma. According to the Bible: the second day (stage) of the creation of the Universe by God.

**Hydrogen plasma** - Highly heated and electrically conductive gas which consists of the split ions of a hydrogen atom (positively charged protons and negatively charged electrons).

**I** - The supreme center of the ideal soul.

**Idea** - The non-physical opposite of matter which possesses no physical attributes or even physical energy.

**Ideal program of physical development** - The complete collection (the meaningful content) of all the laws of nature.

**Ideal space** - The non-physical and infinite opposite of physical and finite space.

**Ideal Universe** - The nonphysical opposite of the physical Universe.

**Ideal World** - The nonphysical opposite of the Physical World.

**Idealism** - A philosophical trend which scientifically proves the primacy of the objective idea and the secondary nature of matter (the scientific opposite of antiscientific materialism).

**Incarnation** - Appearance of soul in the form of living creature.

**Induction** - A method of scientific inference in the course of which from particular initial premises a general law is derived.

**Information** - the ideal (meaningful) content of this or that reality, including physical codes and signals.

**Initial premises (исходные предпосылки)** - The scientifically proven or experimentally confirmed facts, from which, by means of logical deduction, a new judgment can be inferred, termed the result of the proof.

**Inorganic substance (неорганическое вещество)** - substance which possesses no carbon compounds at all.

**Intellect (интеллект)** - The highly developed mind (an ideal category).

**Intelligent (интеллектуал, интеллигент)** - Having intellect

**Intellectual evolution** - The nonrevolutionary process of the lawful ("regular") beginning and development of the intellect on Earth. According to the Bible: The sixth day (stage) of the creation of the Universe by God. ("And God created man in His own image.")

**Ions** - Atoms or groups of atoms which have positive or negative charges caused either by the loss of a part of their electrons or by the addition to them of superfluous electrons.

**Jupiter (Юпитер)** - The fifth planet of the Solar System in order of distance from the Sun.

**Justice (справедливость)** - The harmonious equilibrium of opposing interests.

**Knowledge** - Ideal information which the intellect elaborates, re-shapes, receives, and preserves or transmits.

**Law of Nature** - The obligatory norm of behavior of various physical elements and systems under specific conditions (the particular instructions of the universal program of movement, change, and development of matter).

**Lawful (закономерный)** - Established by law.

**Life** - Conscious existence with many degrees of personal freedom.

**Lifeless matter (неживая материя)** - Physical reality, which possesses one single degree of freedom, which is tied to it from outside by the ideal laws of physical nature. (Lifeless matter is obliged to be in a state of continual movement and change, but only in the way which is prescribed for it by the ideal laws of physical nature; it possesses no mind, no will, and no personal freedom.)

**Lifeless substance (неживое вещество)** - Lifeless matter which possesses volume, weight, and rest mass. According to the Bible: "the dust of the earth."

**Light (scientific conception)** - The photon energy, luminous energy, radiant energy, positive energy.

**Light (biblical conception, 3)** - A positive energy ("And God said, let there be light: and there was light").

**Living creature (живое существо)** - An indivisible and dynamically stable system of automatic self-regulation which possesses physical dimensions, weight, deliberate will, and many degrees of personal freedom.

**Living organism** - The physical form of a living creature.

**Living Substance (живое вещество)** - A physical category which can be governed by genetic codes. Any living substance is obliged to act only in that way which is prescribed for it by its own genetic codes in full accord with the ideal laws of physical nature.

**Love** - The highest level of spiritual delight, which stimulates creation and development.

**Luminary (светило)** - An object, as celestial body, that gives light.

**Magnitude** - The numerical value of a quantity without regard to its sign.

**Mars (Марс)** - The fourth planet of the Solar System in order of distance from the Sun.

**Mass** - The quantitative measure of matter (substance and energy).

**Materialism** - A philosophical trend which groundlessly and without proof proceeds from the totally unfounded assumption of the primacy of matter and the secondary nature of the idea (the antiscientific opposite of scientific idealism).

**Material World** = Physical World.

**Matter** - A physical reality which cannot exist without its internal and external opposite (the zero sum of those real opposites which differ from each other in a fundamental way through some feature, and which negate and at the same time presuppose each other and cannot exist without each other).

**Matter** - An objective reality, possessing physical energy.

**Mercury** (Меркурий) - The planet of the Solar System nearest the Sun.

**Micro-Civilization** - A community of intellectuals who, according to the hypothesis of Academician M. A. Markov, can exist in elementary particles (for example in photons or antiphotons).

**Mind** (ум) - A nonphysical opposite of physical brain, which is able to provide freedom of will consciously.

**Mindless** (неразумный) - Lacking any mind at all.

**Molecule** - The smallest neutral particle of weighable and visible substance, whose further division is not possible without a change in its chemical properties.

**Natural** (закономерный) - Established by the laws of nature, which were created by God.

**Natural Selection** (закономерный отбор) - The particular law of the general program of biological evolution, in accordance with which depending on the conditions of the environment organisms with less perfected genetic codes become extinct and are removed from further circulation, and organisms with more perfected genetic codes survive and experience further development.

Natural selection does not take place spontaneously and independently. Rather, it occurs in a regular fashion, in full accordance with those laws of nature the totality of which constitute the ideal program of biological evolution. Such a perfect program could only be created by the highly intellectual Creator who we call God.

**Neptune** (Нептун) - The eighth planet of the Solar System in order of distance from the Sun, invisible to the naked eye, discovered 1846.

**Neutrino** (нейтрино) - An electrically neutral elementary particle, possessing zero volume, zero rest mass, and a positive half-spin.

**Neutron** (нейтрон) - An electrically neutral elementary particle, whose mass is equal to $1.67 \times 10^{-27}$ kg.

**Nucleons** (нуклоны) - The general name for protons and neutrons.

**Null = Nonexistent** - Amounting to nothing.

**Nullify** - To make null.

**Nullity** - The quality or state of being null.

**Object** (объект) - Something intelligible or cognizable by the mind (the opposite of cognizer).

**Objective** (объективный) - Having actual existence outside of any mind (the opposite of subjective)

**Opposites** (противоположности) = Antipodes, which are equal to each other in magnitude (in quantity) and the reverse in their signs (in quality).

**Organic substance** (органическое вещество) - The substance which contains carbon compounds.

**Organism** = A biological organism.

**Particle** = Elementary particle.

**Perfect vacuum** (абсолютный вакуум) - A physical space in which the quantity of matter is equal to absolute zero (an abstract, virtually impossible category).

**Philosophy** - The science which studies the essence of being.

**Photon** - An elementary portion (quantum) of luminous energy which weight, volume, and total dimensions are equal to zero.

**Photon Plasma** - An unweighable cloud of pure physical energy loaded beyond capacity with photons and having an extremely high temperature: over 10 billion°K. All of the stars and the sun were formed from clouds of photon plasma.

**Physical Energy** - Generalized measure of matter and its physical efficiency (a relativistic determination). Unweighable physical energy has a dual purpose: In the first place it is transformed into a weighable substance; secondly, it brings matter into motion, change, and development.

**Physical Energy** - The general measure of physical work capacity (classical definition).

**Physical Space** - A three-dimensional continuity of unweighable and invisible negative energy, which characterizes the extent and relative disposition of simultaneously existing physical elements and systems in the Universe.

**Physical Universe** = The Universe, where the space is constructed from negative energy, and substance from positive.

**Physical Universe** - The opposite of the ideal Universe.

**Physical World** - The totality of all physical elements and systems without exception. If our Universe is not the only one, then by the Physical World is meant the complex of all physical Universes and anti-Universes. The Physical World is constructed from a zero sum of positive and negative energy, while a zero sum of positive and negative energy is created from nothing.

**Physics** - The science which studies the laws of the motion and change of lifeless matter.

**Planet** (планета) - A nonluminous celestial body, illuminated by light from a star, as the Sun, around which it revolves.

**Planetary Evolution** - The lawful ("natural") process of the separation from primordial clouds of hydrogen plasma of denser fragments, which are later transformed into planets, cooling faster than the entire mass. One of such planets became our Earth. According to the Bible: The third day (stage) of the creation of the Universe by God.

**Plasma** (in physics) - Electrically conductive gas which is heated to the extent that atoms are split into positively and negatively charged ions.

**Plasma** (in biology) - The liquid part of biological structures: blood, lymph, cells (cytoplasm).

**Pluto** (Плутон) - The ninth planet of the Solar System in order of distance from the Sun, invisible to the naked eye, discovered 1930.

**Program** - The complete collection of all instructions or laws having a single purpose (an ideal category).

**Program of Biological Evolution** - An ideal program, created by God and encoded at the level of elementary particles, according to which various forms of plants and living organisms are formed from lifeless matter ("from the dust of the earth").

**Protogalaxy** - A system of protostars or clouds of hydrogen plasma, which inevitably and naturally should be transformed into a galaxy according to the ideal universal program of physical development.

**Proton** - An elementary particle possessing a rest mass of $1.67 \times 10^{-27}$ kg and a positive electrical charge equal in magnitude value to the charge of an electron.

**Protostar** - A cloud of hydrogen plasma which inevitably and naturally should be transformed into a star according to the universal program of physical development.

**Protosun** - A spherical cloud of hydrogen plasma which inevitably and naturally is obliged to be transformed into the sun.

**Psyche** - That ideal content of a living creature which disappears after its physical death (a temporary copy in this world of the otherworldly and eternal soul).

**Quantum** - A discrete portion.

**Quasars** - The nuclei of future galaxies on the periphery of the Universe which are moving away from us at speeds close to the speed of light. The source of energy of the quasars at the present time is cosmic white holes.

**Reality** - Being as compared to non-being.

**Reincarnation** (перевоплощение душ) - The migration of soul from an exhaust dead body into a fresh newborn organism.

**Religion** - The holy faith in the scientifically proven truth of the creation of the Physical World by the Ideal God.

**Rest Mass** - The quantitative measure of a substance contained within a given body which is the same for all inertial systems of calculation.

**Saturn** (Сатурн) - The sixth planet of the Solar System in order of distance from the Sun.

**Scientific Atheism** - The anti-scientific opposite of scientific religion.

**Scientific Religion** - A system of scientific knowledge about the creation of the Relative World by an Absolute God.

**Seeds of Life** - The physical codes of the ideal program of biological evolution, recorded at the energy level inside of unweighable elementary particles such as photons.

**Sex** (секс) - The highest form of physical enjoyment, given by God to the living being for the propagation of the offspring. In addition, sex frees the soul of man from the rust of callous egoism.

**Signal** - The physical means of transmitting and receiving ideal information.

**Solar day** (солнечные сутки) - The time interval during which the Sun completes one full rotation about its own axis.

**Solar System** - A cosmic system consisting of the sun and nine planets (including Earth) which revolve in their own orbits around the Sun.

**Solar year** (солнечный год) - The time interval during which the Solar system completes one full revolution about the center of the Galaxy.

**Soul** - The ideal spirit, which is capable of supporting deliberate life in a physical organism, i.e., that ideal content of a living creature which can live eternally, even after the physical death of the living organism.

**Space** - That in which every relative category, physical or non-physical, exists, moves, and changes.

**Space-Time Continuity** - A united system of space and time (relative category).

**Species** (вид) = **Biological species** (биологический вид).

**Spin** - The natural momentum of the quantity of motion which an elementary particle possesses.

**Spirit** - The ideal and indivisible category of automatic self-regulation which has deliberate will and possesses many degrees of personal freedom. Spirit does not contain within itself any physical attributes.

**Star** - A stable celestial body which radiates light.

**Stellar day** (звёздные сутки) - The time interval during which a star completes one full rotation about its own axis.

**Stellar Evolution** - The non-revolutionary process of the lawful ("natural") conversion of primitive clouds of hydrogen plasma into stars and the sun. According to the Bible: The fourth day (stage) of the creation of the Universe by God.

**Stellar year** (звёздный год) - The time interval during which a star completes one full revolution about the center of its own galaxy.

**Subject** (субъект) = **Cognizer** - A human being who is involved in the cognitive activity of objective reality (the opposite of object).

**Subjective** (субъективный) - of or existing within the mind rather than outside (the opposite of objective).

**Substance** (вещество) - Weighable matter which possesses volume and a positive rest mass (electrons, protons, neutrons, atoms, molecules, and anything which is constructed from them).

**Sun** - An incandescent cosmic sphere which consists mainly of hydrogen plasma, and the star which is closest to us.

**Superstition** - Naive faith in fantastic fables the possibility of which is completely excluded by common sense and the laws of nature. Faith in God is not a superstition because the creation of the Physical World by an Ideal God is validated (and not refuted) by common sense and the laws of nature. Superstition is the atheistic faith in the efficient product of creation without the intellectual Creator, and is in glaring contradiction not only to the laws of nature but also to common sense.

**Tachions** - Elementary antiparticles whose speed exceeds the speed of light in a vacuum.

**Terrestrial day** (земные сутки) = Terrestrial Twenty-Four-Hour Period - The interval of time in the course of which the Earth performs one complete rotation around its own axis.

**Terrestrial year** (земной год) - An interval of time in the course of which the Earth completes one full revolution around the Sun. (One terrestrial year is equal to approximately 365.26 twenty-four hour periods.)

**Time** - The ideal argument which depends on movement, change, and the development of one or another objective reality. (Time is that with whose passage every relative category, physical or non-physical, changes or develops.)

**Tragedy** - The worst form of moral deformity: when falsehood scoffs at truth, madness governs minds, criminals sit in the seats of judgement, and the victims of crime sit in prison.

**Triviality** - Non-being as compared to being.

**Unification** (унификация) - The rational elimination of excessive diversity of elements through the maximum curtailment of their number and the expedient use of similar or identical elements in different systems.

**Universal Program of Physical Development** - The complete combination of all the laws of nature (an ideal category created by God).

**Universe** = **The physical Universe,** unless specifically indicated otherwise - The totality of all those physical elements, systems, planets, stars, and galaxies, without exception, which share a single common physical space.

**Uranus** (Уран) - The seventh planet of the Solar System in order of distance from the Sun.

**Vacuum** - Pure ("empty") physical space, which constitutes a stormy ocean of unweighable and invisible negative energy (the physical category). According to the Bible: "the darkness."

**Venus (Венера)** - The second planet of the Solar System in order of distance from the Sun.

**Water (H₂O)** - The most common liquid substance on Earth, representing a compound of hydrogen and oxygen. Nearly 75% of the earth's surface is covered by water, forming seas, oceans, rivers and lakes. However, the word "water" in the Bible expresses not only the concept of water as $H_2O$, but also hydrogen ($H_2$), hydrogen plasma, and the hydrogen cloud. The expression "water that is under the heavens" means that cloud of hydrogen plasma from which our Solar System was formed. The expression "water that is above the heavens" means that cloud of hydrogen plasma from which all the other stars and galaxies were formed.

**Weight** - The product of rest mass and the acceleration of the free fall of a physical body.

**White Cosmic Hole** - An ideal point, at which an equal quantity of positive and negative energy arises from nothing continuously and simultaneously.

**Will (воля)** - The purposefulness of the ideal spirit.

# SUBJECT INDEX

\* g = Glossary

Universe 72,187,293g
  age 85,126,134,243
  birth 85,112,125f,243
  density 86
  dimensions 90
  evolutionary 105,135
  expansion 81,88
  formation 133
  genesis 91,111,162
  ideal 72
  origin 91,112,127
  physical 72
  stationary 106,135
Uranus 125f,161,177,179,293g

Vacuum 69,73,128,293g
Venus 125f,161,177,179,294g
Volume 33
Water 22,162,295g
Weight 28
Weightless (невесомый) 28,44,243
Weighty (весомый) 28,44
Where God is 100,102
White hole 129,134,294g
Who created God 94,209,219,294g
Will 94,209,219,294g
World 21,72,77,91
Zero 49-56,127
Zero sum 49-56

# NAME INDEX

Adam and Eve 125f,230-1,241
Alfven 158,161
Arrhenius, Svante 164
Bacon, Francis (1561-1626) 82
Bernal, John (1901-1972) 218
Bohr, Niels (1885-1962) 11
Cocconi 167
Communist Party of the Soviet
Union 25-6,186
Copernicus, Nicholas (1473-1543)
164
Darwin, Charles (1809-1882) 210
Davydov, Joseph 7,176
Dirac, Paul 39,51,69,127
Doppler 84
Eigen, Manfred 167
Einstein, Albert 14,27,33,58,66,82-
4,92,152,249-50,265
Engels, Friedrich (1820-1895) 272
Fridman, Alexander 84,167
Glazer, Rolland 203
Godunov, Boris 236
Golin, Genrikh 11
Golubev, Daniel 264
Gorbachev, Michael 25
Hoyle, Fred 106,162
Ivanov, Sergey 273
Kamenetskiy, Saveliy 274
Kant, Immanuel (1724-1804) 158,
164
Kapitsa 233
Kardashev, N.S. 72,175
Kashchey Immortal 251
Kitaygorodskiy, Alexander 84
Koulechov, Guennadi 266
Laplace, Pierre 158,162

Lenin, Vladimir (1870-1924) 56,62,103,
277
Levitan, Ephraim 183
Lorentz, Hendrik 152
Markov, M A. 35,167
Max Ernst (1838-1916) 54
Messiah = Moshiah 242
Milne 82
Moses 19,121,242,249
Novikov, I. D. 144
Oparin, A. 167,202,215,218,252
Pasteur, Louis (1822-1895) 218
Pekelis, Viktor 215
Planck, Max (1858-1947) 38
Poincaré, Henri (1854-912) 11,45,54,70
Ponnameruma, Cyril 15
Pushkin, Alexander 251
Reznitskiy, Yevgeniy 209
Rydnik, Vitaliy 38,68,253
Sagan, Carl 163
Shakespeare 219
Shklovsky, Joseph 81,85-6,105,132,138,
162,166,192,218
Shoul ben Matthew 5
Sinaysky, G. 23
Stalin, Joseph 10,15,257
Su Hsiu Huang 160
Vyshkina, Tamara 267
Yaroslavskiy, Yemelyan 137-8,167,236,
238
Zeldovich, Ya. B. 127
Zhdanov, A. A. 81